Assurance Technologies

McGraw-Hill Engineering and Technology Management Series
Michael K. Badawy, Ph.D., Editor in Chief

Assurance Technologies

Principles and Practices

Dev G. Raheja
System Assurance Consultant
Lecturer, Graduate School
University of Maryland

McGraw-Hill, Inc.

New York   St. Louis   San Francisco   Auckland   Bogotá
Caracas   Hamburg   Lisbon   London   Madrid
Mexico   Milan   Montreal   New Delhi   Paris
San Juan   São Paulo   Singapore
Sydney   Tokyo   Toronto

*Dedicated to the ideals of people like Mother Teresa
and Mahatma Gandhi, who remind us that
technology must contribute to lifting mankind from
misery.*

Library of Congress Cataloging-in-Publication Data

Raheja, Dev.
  Assurance technologies : principles and practices.

    p. cm.
  Includes index.
  1. Quality assurance. 2. Design, Industrial. 3. Reliability
(Engineering) I. Title.
TS156.6.R34 1991   658.5'62—dc20   90-39147
ISBN 0-07-051212-4

1 2 3 4 5 6 7 8 9 0 DOC/DOC 9 6 5 4 3 2 1

ISBN 0-07-051212-4

*The editors for this book were Robert W. Hauserman, Caroline Levine,
and George Watson. The designer was Naomi Auerbach and the
production supervisor was Suzanne W. Babeuf. It was set in Century
Schoolbook by McGraw-Hill's Professional Publishing composition
unit.*

Printed and bound by R. R. Donnelley & Sons Company

*For more information about other McGraw-Hill materials,
call 1-800-2-MCGRAW in the United States. In other
countries, call your nearest McGraw-Hill office.*

Contents

Series Introduction

Technology is a key resource of profound importance for corporate profitability and growth. It also has enormous significance for the well-being of national economies as well as international competitiveness. Effective management of technology links engineering, science, and management disciplines to address the issues involved in the planning, development, and implementation of technological capabilities to shape and accomplish the strategic and operational objectives of an organization.

Management of technology involves the handling of technical activities in a broad spectrum of functional areas including basic research; applied research; development; design; construction, manufacturing, or operations; testing; maintenance; and technology transfer. In this sense, the concept of technology management is quite broad, since it covers not only R&D but also the management of product and process technologies. Viewed from that perspective, the management of technology is actually the practice of integrating technology strategy with business strategy in the company. This integration requires the deliberate coordination of the research, production, and service functions with the marketing, finance, and human resource functions of the firm.

That task calls for new managerial skills, techniques, styles, and ways of thinking. Providing executives, managers, and technical professionals with a systematic source of information to enable them to develop their knowledge and skills in managing technology is the challenge undertaken by this book series. The series will embody concise and practical treatments of specific topics within the broad area of engineering and technology management. The primary aim of the series is to provide a set of principles, concepts, tools, and techniques for those who wish to enhance their managerial skills and realize their potentials.

The series will provide readers with the information they must have and the skills they must acquire in order to sharpen their managerial performance and advance their careers. Authors contributing to the series are carefully selected for their expertise and experience. Although the series books will vary in subject matter as well as approach, one major feature will be common to all of them: a blend of practical applications and hands-on techniques supported by sound research and relevant theory.

The target audience for the series is quite broad. It includes engineers, scientists, and other technical professionals making the transition to management; entrepreneurs; technical managers and supervisors; upper-level executives; directors of engineering; people in R&D and other technology-related activities; corporate technical development managers and executives; continuing management education specialists; and students in engineering and technology management programs and related fields.

We hope that this series will become a primary source of information on the management of technology for practitioners, researchers, consultants, and students, and that it will help them become better managers and pursue the most rewarding professional careers.

Michael K. Badawy

PROFESSOR OF MANAGEMENT OF TECHNOLOGY
THE R. B. PAMPLIN COLLEGE OF BUSINESS
VIRGINIA POLYTECHNIC INSTITUTE AND STATE UNIVERSITY
FALLS CHURCH, VIRGINIA

Preface

This book is written for design engineers who have a strong urge to be familiar with product assurance technologies. The book is unique in that it covers all the major technologies in one volume. Each chapter introduces an assurance technology and includes a liberal sampling of case histories and examples from the author's experiences of more than 25 years.

The book is directed to design engineers at any level of experience in any field. It is a source of practical applications for graduate students, but it is written with sufficient depth that the practicing design engineer will find it easy to team up with realiability, maintainability, safety, quality, and logistics engineers. If this objective is met, the author has succeeded in his mission. Where details are covered adequately by other books, such references have been provided.

A major strength of this book is that it will help bridge the gap between academia and the industry. This gap today is wide and has been the largest contributor to misapplication of assurance principles.

The author has trained over 20,000 engineers and managers worldwide through short courses at several universities and corporations in the consumer, industrial, and government sectors. A predecessor of the book has been used at the Graduate School of the University of Maryland in its master's program in reliability engineering. To bridge the gap between universities and the industry, each chapter contains topics for student projects and master's theses. An instructor can create more topics easily with the help of this book.

The author wishes to emphasize the message of industry leaders who strongly urge that this knowledge be imparted to all engineers. Most engineering programs hardly scratch the surface. They must teach how to use assurance technologies in making products better, faster, and at lower cost.

Dev. G. Raheja

Acknowledgments

I wish to thank the following corporations for providing valuable examples: McElrath & Associates, Delco Electronics (General Motors), Ford Motor Co., Boeing, Intel, and Yokogawa-Hewlett Packard. I want to express my deep appreciation to my colleagues, who provided valuable comments and help. In particular I wish to mention Dr. Wayne Nelson, Melvin Chambers, Julius Wang, Richard Rudy, Dr. Harrison Wadsworth, Dr. Mohammad Modarres, Dr. Al Szews, Alan Golant, Dean Jones, Nanik Lahori, Gita Raheja, Ajay Chamania, my editors, and my wife Hemlata, who typed the entire manuscript.

Impact of Assurance Technologies on Profits

1.1 Introduction

Assurance technologies are processes for assuring that a product performs well during its lifetime. They call for proactive care in design, manufacture, and maintenance, and they must be integrated if they are to succeed. Manufacturers of large systems such as aircraft or nuclear plants call the integration process *system assurance*. Manufacturers of consumer or industrial products call it *product assurance*. The major difference lies in the degree of complexity.

The basic assurance technologies are:

Reliability analysis: Lowers product failure rate over the long term, reduces producer's warranty costs, and makes the customer want to come back—if the price is not unreasonable. Product reliability depends on design, component quality, and manufacturing. Its design requirements come from preliminary analysis before any hardware or software is acquired.

Maintainability analysis: Lowers downtime, reduces repair time, and, as a result, reduces maintenance costs. Some of the inputs for maintainability come from reliability analysis.

System safety analysis: Eliminates the causes of safety-related incidents from misuse, human factors, and design oversights.

Quality assurance engineering: Incorporates customer satisfaction requirements in design and controls manufacturing defects.

Human factors engineering: Recognizes the role of humans in product malfunctions. Helps designers prevent human-induced mishaps by making designs insensitive to human errors of use.

Logistic engineering: Reduces ongoing field support costs, most of which result from poor quality, reliability, maintainability, and safety. A design analysis at the outset helps avoid much of this unnecessary waste of resources.

1.2   Cheaper, Better, and Faster Products

The purpose of assurance technologies is to allow a manufacturer to make products better, cheaper, and faster. These benefits are the best leverage a manufacturer has in a competitive market.

The U.S. Department of Defense numbers on life-cycle costs[1] show that 85 percent of costs are incurred during production and in service and only 15 percent during design and development. The life-cycle cost breakdown is as follows:

Concept stage	3%
Development stage	12%
Production stage	35%
Operation and support stage	50%

If the life-cycle costs (LCC) are $100 million for a product, $50 million dollars are spent in the field. Even though most of these costs are paid by the customer, they have tremendous influence on the supplier. The customer can always go to another supplier whose product requires less maintenance and repair (see Chap. 4 for an example; it tells how Canon took away a 90 percent share of a market from a competitor). Even an old, established supplier may not survive when this happens. Aside from inconvenience and downtime, a high maintenance and repair rate means higher product cost, and a supplier whose total product cost is too high inevitably starts losing customers. Moreover, the 35 percent contribution from the production stage includes tremendous waste in the form of rework and lost yield. However, with assurance technologies tools such as process reliability, the cost of production waste can be drastically reduced.

A cost reduction effort requires some investment in engineering analyses and in improving the product and process design and process controls. There is enough evidence to show that investment in these technologies is not an add-on cost, but a smart investment. Companies that are not willing to invest $1 today to save $10 later (proprietary data show as much as $100) are likely to stagnate.

Figure 1.1 makes the benefits of assurance technologies clearer. It shows that life-cycle costs are highest when only a small investment in assurance technologies is made and become lower and lower as more technologies are integrated. For example, an LCC of $100 mil-

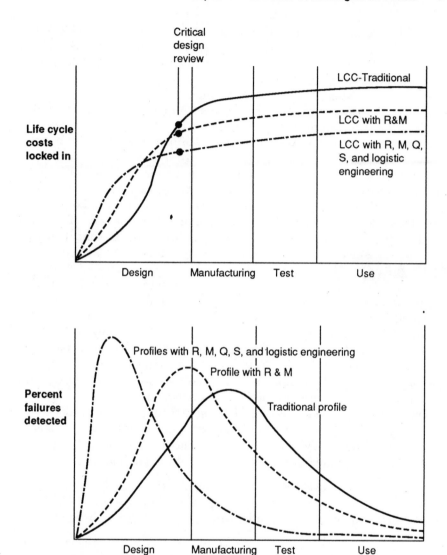

Figure 1.1 Effects of assurance technologies on life-cycle costs and performance.

lion can be reduced to $50 million with assurance technologies. The benefits go to both the supplier and the customer (see the discussion of LCC with and without screening in Chap. 3). A company should not be afraid to invest $5 million to save $50 million in future costs—except if it is run only for short-term survival and quick profits.

So far the benefits of low cost and good performance have been stressed. Fortunately, these benefits are also the primary requisites for adhering to an efficient implementation schedule. The *Challenger* space shuttle disaster taught lessons in this regard. The accident in 1986 showed that space shuttle schedules were totally out of hand because of technical problems. Schedules can be held to only if they are realistic and are backed by sufficient resources to identify trouble spots and take steps to resolve them. Successful schedule control depends on the performance of the system and control of the demands made of it, both tangible and intangible. The just-in-time system of inventory control is an excellent example of how schedules can be enhanced through quality and reliability, with massive cost reductions.

The product introduction cycle time can also be reduced with *concurrent engineering*. This process allows many kinds of analyses simultaneously so that problems are prevented downstream. For example, an associate of the author made a reliability analysis of a new product for a defense electronics firm in Minneapolis concurrently with product design. He showed that many components were operating in the upper 50 percentile range of their ratings in the current design. He convinced the company that this design was not in conformity with reliability engineering principles, and he helped redesign the product so that all the parts operated below the 50 percent rating. The result took design engineers by surprise. When they tested the product for qualification, the product passed! They had never had this experience before; they had always gone back and forth several times, testing and fixing the problems. This time they were 3 months ahead of schedule because the new product did not require fixing.

This example shows a result of reliability improvement alone. If similar improvements are also made in quality assurance, system safety, maintainability, and logistics support engineering, new product introduction schedules should be shortened even more.

Caution: The key is concurrent engineering. If the analyses are done sequentially, as they are in many companies, they will always be late and of little value, and schedules will be almost always beyond control.

1.3 What Is System Assurance?

To understand system assurance, one has to understand the definition of a failure. If a system does not meet the reasonable expectation of the user, then it has failed, even though it meets the specifications. This statement will surprise many readers, but millions of unsafe products are recalled annually for this reason. An example: a man wearing a programmable pacemaker walked through a department store's electromagnetic antitheft device. The device caused the pacemaker to malfunction; the man died of a heart attack as a result. The pacemaker and the antitheft device met their specifications, but did not meet the expectations of their users. A customer expects a safe product.

1.4 Key Management Responsibilities

1.4.1 Integration

For a customer, the system performance depends on at least four elements that only supplier management can integrate: hardware, software, people, and methods. The supplier's system should be robust against errors in these areas in the user's environment. The user expects this robustness throughout the entire life cycle. These concepts are shown in Fig. 1.2.

The block diagram in Fig. 1.2a shows that, for a system to work, the hardware, the software, the manufacturing methods, and the human factor (operating and maintenance personnel) must be reliable simultaneously. Also, implied in the model is that all of the interfaces among these four elements must be reliable.

Since human behavior is the most unpredictable among the elements, it may be the weakest link in the chain. Figure 1.2b shows an improved model in which the human user is a redundant element. With increasing use of software, the system needs less human interaction.

1.4.2 Budget consistent with objectives

Once a system is defined, management must observe the ground rules and manage resources effectively. Resources are needed to finance highly detailed analysis during early design. An investment is most productive at this point. Unfortunately, the lack of a budget prevents

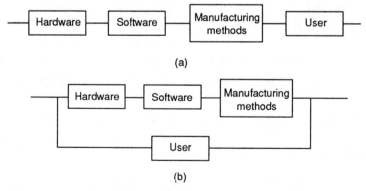

Figure 1.2   (a) A typical system assurance model. (b) An improved model.

managers from putting investment dollars on the table. To develop budgets for such activities, managers should develop program plans that include tasks, sequences, and milestones. As much as possible, the tasks should be performed by parallel groups.

1.4.3 Managing risk

The process of risk management is very crudely defined in most management texts. Most of them go by past experience. They do not take into account that there is always a first-time disaster waiting to happen. This book approaches risk in a way designed to avoid many first-time disasters with the help of proactive designs.

The classical definition of risk as "a product of frequency and severity" is vague. It does not indicate what level of severity should be considered if the product has, say, 10 levels of severity. It does not specify which out of, say, 5000 or so failures should be considered when frequency of failure is computed. Harold Geenan, former chairman of ITT Corp., writes "Theory should be considered a starting point. Once you have broken that barrier you have to expand into the reality."[2] This definitely applies to risk. The first reality is that a complex product, the *Challenger* shuttle, for example, can fail in several thousand ways. Each failure can have several levels of severity from no harm to loss of the system and loss of life. Indeed, the *Challenger* disaster may have been even more severe—it might have crashed in a crowded section of a city. A second reality is that a complex product such as the shuttle can fail again in a different manner.

The only way to manage risk is to anticipate what can go wrong and make timely attempts to design the product to overcome failures. The risk management process, therefore, must consist of the following steps:

1. Identify the failures likely to occur by proactive analyses such as design reviews; preliminary hazard analysis; and failure-mode, effects, and criticality analysis.

2. Develop alternatives to control the failures. Control by design is usually cheaper than testing products.

3. Implement the acceptable solution without creating additional risks.

4. Control the risk through a closed-loop system.

1.5 Is System Assurance a Process?

All assurance technologies are processes. Thus, system assurance is a process. However, like statistical process control, it will only work if

(a)

(c)

Figure 1.3 (a) Annual failure rate vs. list price. (b) YHP product-line achievements. (c) Reliability improvement. (*From a presentation by Kenzo Sasaoka, President, Yokogawa-- Hewlett Packard, to the International Symposium on Reliability and Maintainability, organized by the Japanese Union of Scientists and Engineers, 1990.* Reprinted with permission of Mr. Sasaoka) top management treats it as a business goal. Management not only must be committed to system assurance, but also should be actively involved. Management that looks only at financial results will no longer be able to compete.

Example 1.1 A classic example of the above is Yokogawa-Hewlett Packard where management as well as all the research and development (R&D) engineers learned and practiced reliability engineering. The data in Fig. 1.3 a shows the annual failure rate versus the list price. As it should be, the reduction in price is accompanied by the corresponding reduction in failure rate. If assurance technologies are used effectively, the low failure rate should result in higher profits, as shown in Fig. 1.3 b. In the 1980s John Young, President and Chief Executive Officer of Hewlett Packard, set the companywide goal of tenfold improvement in reliability. Figure 1.3 c is proof that such goals have been achieved.

In summary it is fair to say that assurance technologies reduce cost and increase profits. As this book will show later, they also reduce the cycle time for introducing new products.

References

1. M. B. Darch, "Design Review," in *Reliability and Maintainability of Electronic Systems*, J. E. Arsenault and J. A. Roberts (eds.), Computer Science Press, Potomac, Md., 1980.
2. Harold Geenan, *Managing*, Doubleday, New York, 1984.

Further Reading

Ireson, Grant W., and Clyde F. Coombs, Jr., *Handbook of Reliability Engineering and Management*, McGraw-Hill, New York, 1988.
Ohmae, Kenichi, *The Mind of the Strategist*, McGraw-Hill, New York, 1982.

Introduction to
Statistical Concepts

2.1 Probabilistic Designs

This chapter covers statistics only to the extent necessary for understanding the other chapters in the book. The reader is advised to read Refs. 1, 2, and 3 for more details.

Performance of most designs is treated as deterministic, but it is always probabilistic. If an engineer uses Young's law to calculate the strain for the stress applied to each of a group of similar components, the law will give the same value for each. In real life, the strain for each component will not be same. The range of strains can be described with statistical theory. The same variability principles apply to all engineering measurements and values. There is always a variation, and the range of the variation can be quantified.

To study variation, we draw a histogram of data and determine what percentage is below or above a target value. When an instrument supplier promises an accuracy of ±5 percent, it means a very high percentage of measurements will fall within ±5 percent around a target measurement.

If histograms are plotted so that each bar represents the frequency as a percentage of all the values in the population, and if each interval is small, the heights of the bars can be joined to form a smooth curve. This curve represents the relative frequency with which the events represented by the interval occur. The relative frequency, when used as an estimate of the population fraction, is commonly referred to as the *probability*. The mathematical model of this curve is called the *probability density function* (pdf) or $f(t)$.[1,2]

This chapter covers some distributions that are applicable to engineering situations. Special distributions may be required if these models do not match the data. Section 2.2 covers the generic construction of a probability density function. Other sections describe the models frequently used in assurance technologies.

2.2 Probability Computations for Reliability, Safety, and Maintainability

Probability of failure is the single most important concept encountered in reliability, maintainability, and safety assessment. The following generic concepts will be very helpful, no matter what the shape of the distribution.

Construction of a histogram and the empirical distribution. A probability distribution is determined from a histogram of the data. Even though the future distribution may not be the same, a histogram is an estimate which can be modified with the accumulation of more knowledge in the future. Constructing the histogram involves grouping the data in mutually exclusive and collectively exhaustive intervals, computing the sample fraction for each interval, and plotting fraction versus the interval. *Mutually exclusive* means no two intervals will overlap. *Collectively exhaustive* means we must consider all the intervals that are possible even though there have been no data in some intervals. For example, the following liability claims were made during a 10-year period:

Size of claims, dollars	Number of claims
0– 4,999	13
5,000– 9,999	31
10,000–14,999	14
15,000–19,999	11
20,000–24,999	5
25,000–29,999	2
30,000–34,999	0
35,000–39,999	1

Each interval in the above data is already arranged so that there is no overlapping. The next step is to determine if the intervals are collectively exhaustive. Since there can be no negative claims, the claims should start from 0. Up to $40,000, all possibilities are accounted for. Can a claim exceed $40,000 even though records do not show that any has in the past? It is possible; future claims might be very large. So one more interval is necessary. It can be labeled $40,000 + to include all possible claims beyond $40,000. Even though the sample shows the observed fraction for $40,000 + claims as zero, a larger sample may reveal some small positive value. The observed probability for each interval is computed by dividing the number of claims in each category by the total number of claims, as tabulated below.

Size of claims, dollars	Number of claims	Observed probability, %	Cumulative probability, %
0– 4,999	13	16.9	16.9
5,000– 9,999	31	40.2	57.1
10,000–14,999	14	18.2	75.3
15,000–19,999	11	14.3	89.6
20,000–24,999	5	6.5	96.1
25,000–29,999	2	2.6	98.7
30,000–34,999	0	0	98.7
35,000–39,999	1	1.3	100.0
40,000 +	0	0	100.0
Total	77	100.0	

Figure 2.1 depicts the observed probabilities in a histogram (shown in smaller intervals). Even more useful are the cumulative probabilities plotted in Fig. 2.2.

The cumulative distribution plot can be used for reading various risks at a glance. For instance, what are the chances (probability) that the next claim size will be below $22,000? The answer is obtained simply by projecting a vertical line from $22,000 until it intersects the plotted curve and reading the y value. The observed probability for this example is 93 percent.

Caution: Some statisticians recommend that the Bayes approach be used in estimating probability. This approach combines past experi-

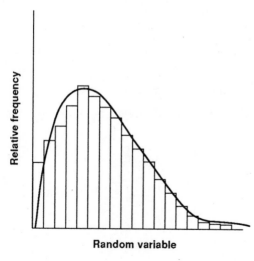

Figure 2.1   Construction of probability distribution.

Relative frequency

Random variable

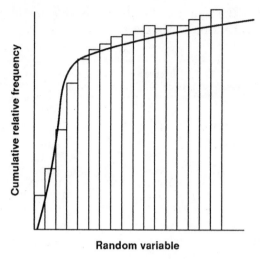

Cumulative relative frequency

Random variable

Figure 2.2 Construction of cumulative probability distribution function.

ence and new data to generate a new distribution. Readers with adequate knowledge of statistics should take this into account.

For an entirely new product or a system on which no past data are available, the failure modes are first determined by means of logic models, as in fault-tree analysis. The probabilities for various failures may be assigned by using expert opinions. The probability of an accident may be computed by using the rules explained in Chap. 5.

Computing reliability. The reliability of a product is derived from the following relations:

Probability of product working + probability of product failure = 1

Therefore,

Probability of product working = 1 − probability of product failure

Or

Reliability = 1 − probability of product failure

Two types of reliability computations are generally required (Fig. 2.3):

1. Reliability up to time t is

$$R(t) = 1 - \text{probability of failure up to time } t$$

$$= 1 - \int_0^t f(t)\, dt = 1 - F(t)$$

(a)

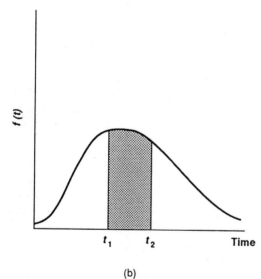

(b)

Figure 2.3   (a) Reliability at time t;
(b) reliability during an interval.

where $F(t)$ is the cumulative probability up to time t, also called cumulative distribution function (cdf).

2. Reliability during an interval $(t_2 - t_1)$ is defined as

$$R(t_2 - t_1) = 1 - \int_{t_1}^{t_2} f(t)\, dt$$

Sometimes the probability of failure during an interval can be defined in terms of reliability. For example,

$$P(t_1, t_2) = F(t_2) - F(t_1)$$

$$= [1 - R(t_2)] - [1 - R(t_1)]$$

$$= R(t_1) - R(t_2)$$

Failure rate and hazard function. The interval failure rate during a time interval (t_2, t_1) is defined as the fraction (or percentage) of units failing per unit time (or distance or cycles). That is, in the interval, given that these units have not failed prior to time t_1,

$$\text{Average interval failure rate} = \frac{N_1 - N_2}{N_1(t_1 - t_2)} \qquad (2.1)$$

where N_1 = number of survivors at time t_1
N_2 = number of survivors at time t_2
$N_1 - N_2$ = number failed during the interval (t_2, t_1)

The *hazard function* is defined as the limit of the failure rate as $(t_2 - t_1)$ approaches zero. It is also called the *instantaneous failure rate*. That is, when the interval $(t_2 - t_1)$ is very small, the failure rate is denoted by the hazard rate which can be written as $h(t) = f(t)/[1 - f(t)]$ or $f(t)/R(t)$.[1]

If one desires a probability of failure in a small interval, from t to $(t + \Delta t)$, the product $h(t) \Delta t$ is the probability. The hazard rate varies over time for most distributions. For an exponential distribution (Sec. 2.5), the hazard rate is constant.

2.3 Normal Distribution

The normal distribution model typifies the concept of variation. It adequately describes many quality control measurements, test data, and wearout failures for mechanical components. Some electronics characteristics also are adequately described by the normal distribution. Ideally, the products of a process will have exactly the same measurements, but this does not happen. The measurements will tend to crowd around a value. This value in the normal density function of Fig. 2.4 is represented by the *arithmetic average*, called $\overline{X}$. It is used as an estimate of the population mean μ. The data are distributed symmetrically around the population average. In other statistical distributions, the data may not be spread symmetrically.

Consider a dimension on a part. All pieces do not measure exactly the same, and their departure from the average is called deviation.

Parameter value

Figure 2.4   Normal probability density function.

This gives a measure of spread, which can be written as $(X_i - \overline{X})$ where X_i is the individual measurement and $\overline{X}$ is the average of all the measurements. The average deviation in this case cannot be computed arithmetically. The positive and negative deviations sum to zero. Thus the average deviation is not a suitable measurement of spread. Statisticians, however, have a way to compute spread; they square each deviation, making it always a positive number. If a random sample of n units is drawn from a lot, the average of the squared deviation then is

$$\frac{\sum (X_i - \overline{X})^2}{n}$$

where $\overline{X}$ is the arithmetic average of all the measurements in the sample. This positive value allows computation of the average deviation by taking the square root of this value. The average deviation computed this way is called *standard deviation*. For a sample, it is written as

$$s = \sqrt{\frac{\Sigma (X_i - \overline{X})^2}{n}} \qquad (2.2)$$

For an unbiased estimate (see Ref. 2) of population when the sample size is less than 30, the formula is written as

$$s = \hat{\sigma} \sqrt{\frac{\Sigma(X_i - \overline{X})^2}{n - 1}}$$

A property of the normal probability density function is that 67 percent of the population is within ± 1 standard deviation of μ, 95 percent is within ± 2 standard deviations of μ, and 99.7 percent within ± 3 standard deviations (Fig. 2.5).

The terms $\overline{X}$ and s usually refer to statistics computed from the sample data. Such values for population (called *parameters*) are represented by the Greek letters μ and σ.

The pdf for the normal distribution is

$$f(t) = \frac{1}{\sqrt{2\pi}\sigma} \exp\left[-\frac{(t - \mu)^2}{2\sigma} \right]$$

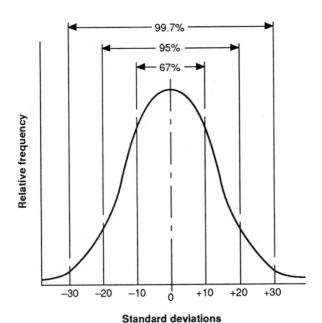

Standard deviations

Figure 2.5   Percent area under the normal density curve.

where t is the measurement under consideration, called the *random variable*.

Table A in the appendix provides the numerical value of the area under the normal density curve for a given value of Z. It can be used to calculate the population percentage above or below a selected value of the variable. Typically an engineer wants to estimate what percent of production is expected to be within or outside a desired specification. The only information needed is a score called Z, determined from the sample average $\overline{X}$ and the number of standard deviations between $\overline{X}$ and the specification:

$$Z = \frac{\text{specification} - \overline{X}}{\text{standard deviation}} \qquad (2.3)$$

With this score one can look up the percentage outside the specification in the table. The computation is also done for each side of the specification, if the specification is two-sided. The procedure is demonstrated in the following examples. The examples assume the data are normally distributed. The analyst should verify this assumption either analytically or graphically. See Ref. 4 for verification procedures.

Example 2.1 The data below were collected on a random sample from a production batch. The specification for the Rockwell hardness number is the range 54 to 58.

Rockwell hardness numbers			
54	58	57	56
56	55	54	55
57	56	57	55
56	56	55	56
57	54	57	55

Compute:

1. The process capability defined as ± 3 standard deviations spread. (*Note*: A recent trend is to use ± 4 standard deviations.)
2. The percent production expected to be out of specification.
3. The ppm level (nonconforming parts per million).

solution

X_i	$(X_i - \overline{X})^2$
54	3.24
56	0.04
57	1.44
56	0.04
57	1.44
58	4.84
55	0.64
56	0.04
56	0.04
54	3.24
57	1.44
54	3.24
55	0.64
57	1.44
56	0.04
55	0.64
55	0.64
56	0.04
55	0.64
57	1.44
1116	$\Sigma(X_i - X)^2 = 25.20$

$$\overline{X} = \frac{1116}{20} = 55.8$$

The estimate s of population standard deviation is

$$\sqrt{\frac{\Sigma(X_i - \bar{x})^2}{n - 1}} = \sqrt{\frac{25.2}{20 - 1}} = \sqrt{1.33} = 1.15$$

Then

$$\text{Process capability} = 55.8 \pm 3 \text{ standard deviations}$$

$$= 55.8 \pm 3 \times 1.15$$

$$= 55.8 \pm 3.45$$

This suggests that the process spread is expected to be in the range of 52.35 to 59.25 for 99.7 percent of the production (see Fig. 2.5). Only 3 out of 1000 are expected to be out of this range.

To estimate the percent production expected to be out of specification, compute the percentage outside each end of the specification. For the upper side of the specification, from Eq. (2.3),

$$Z = \frac{|58 - 55.8|}{1.15} = 1.91$$

The area outside the specification for Z = 1.91, from Table A in the Appendix, is 0.0281 or 2.81 percent.

For the lower side of the specification:

$$Z = \frac{|54 - 55.8|}{1.15} = 1.57$$

The area outside the lower side from Table A for Z = 1.57 is 0.0582 or 5.82 percent. This process is therefore inherently expected to produce 2.81 + 5.82 = 8.63 or approximately 9 percent units out of specification.

The out-of-specification parts in parts per million units of production are

$$\frac{8.63}{100} \times 1{,}000{,}000 = 86{,}300 \text{ ppm}$$

Example 2.2 Consider the pacemaker manufacturer that told doctors that the average battery life is 4 years and the doctors who told the patients that their batteries will work for 4 years. Assume normal distribution estimate:

1. The number of patients whose batteries may fail in 4 years.
2. If the standard deviation of battery life is 5.8 months, what proportion of batteries is expected to fail in the first 3 years?

solution For a normal distribution, 50 percent of the distribution lies on each side of the average (Fig. 2.6). Therefore, 50 percent of the patients are expected to see failures of the battery by 4 years. The doctors gave erroneous information to the patients that their batteries will last 4 years. Only half the patients are expected to find this statement true.

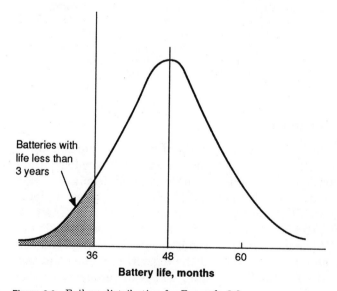

Figure 2.6 Failure distribution for Example 2.2.

Note that the word *average* can be misleading. Many users think it means the minimum guaranteed value. When a light bulb bought from a store is said to have an average life of 2000 hours, many people expect every light bulb to last 2000 hours or more. They do not realize that half of them will fail before 2000 hours.

The proportion of batteries expected to fail in the first 3 years is found from Eq. (2.3):

$$Z = \frac{|36 - 48|}{5.8} = 2.07$$

From Table A in the appendix, the proportion of batteries failing in the first 3 years, for $Z = 2.07$, is 0.0192, or 2 percent rounded off.

Example 2.3   A specification requires that a device activate when a stress of 7 lb/in^2 or more is applied to it. The following data on activation stress in pounds per square inch were collected on 34 devices out of a lot of 300 devices. The data show that all devices were activated within the design limits. Is this design safe?

22.5	7.5	17.5	17.5	22.5
32.5	40.0	35.0	42.5	17.5
27.5	32.5	47.5	15.0	15.0
10.0	15.0	12.5	32.5	15.0
15.0	25.0	22.5	22.5	12.5
27.5	30.0	20.0	15.0	27.5
25.0	17.5	22.5	10.0	

solution   This device may not be safe. Even though all the values are within specification, the population of remaining untested devices $(300 - 32 = 268)$ is expected to have some defective devices. The estimates computed from the sample are $\overline{X} = 22.65$ and $s = 9.71$. Z for the 7.0-lb/in^2 specification is $(22.65 - 7.0)/9.71$ or 1.61. The fraction defective from appendix Table A is 0.0537 or 5.37 percent. Therefore, $268 \times 0.0537 = 14$ devices can be unsafe.

2.4   Log Normal Distribution

In many applications, especially in reliability and maintainability, the data may not fit the normal distribution. However, the logarithms of the random variable may readily fit the normal distribution, hence the name *log normal*. If the log normal distribution applies, a plot of the data on log normal graph paper will be a straight line. The plotting procedure for this distribution and others is the same; it is discussed in Sec. 2.7. The analytical procedure consists of computing the average and the standard deviation of the logarithm values and taking the antilogarithms of the final results. This will become clear with the following example.

Example 2.4 The data below were collected for type 7075.T6 clad aluminum tested at 12,000 lb/in^2. The aluminum was tested in the lab and in the field.

	Field results			Laboratory results	
i	Probability, %	Life in cycles	i	Probability, %	Life in cycles
1	2.08	277,000	1	2.0	212,500
2	6.25	303,600	2	6.0	234,800
3	10.42	311,800	3	10.0	249,500
4	14.58	321,500	4	14.0	253,300
5	18.75	322,800	5	18.0	256,900
6	22.92	332,200	6	22.0	259,400
7	27.08	337,500	7	26.0	260,400
8	31.25	339,400	8	30.0	260,900
9	35.42	355,000	9	34.0	265,500
10	39.58	364,000	10	38.0	265,500
11	43.75	372,200	11	42.0	266,400
12	47.92	384,600	12	46.0	268,200
13	52.08	396,300	13	50.0	269,800
14	56.25	405,800	14	54.0	272,600
15	60.42	430,900	15	58.0	279,700
16	64.58	431,000	16	62.0	279,900
17	68.75	439,600	17	66.0	286,300
18	72.92	505,500	18	70.0	292,000
19	77.08	518,700	19	74.0	299,500
20	81.25	519,900	20	78.0	319,900
21	85.42	547,200	21	82.0	324,600
22	89.58	579,200	22	86.0	327,800
23	93.75	643,100	23	90.0	329,100
24	97.92	887,400	24	94.0	364,800
			25	98.0	367,300

1. Calculate the number of cycles at which the chance of failure is less than 5.0 percent. Assume a log normal distribution. (This information will be used in planning preventive maintenance).
2. Plot the data on log normal graph paper to make sure the model is valid. Calculate the life multiplication factor if the laboratory results are to be used to predict field results. The failure mode in the lab is identical to that in the field. Is the log normal model valid? Is the prediction valid?

solution Converting the field life-cycle data to logarithms (base 10) gives

$$\overline{X} = 5.4477$$
$$s = 0.0566$$

For the chance of failure to be less than 5.0 percent, the value of Z from Table A in the appendix is 1.65 (for area .0495). The cycles to failure at 5 percent probability (the specification in this case) can be calculated from the value of Z:

$$Z = 1.65 = \frac{\overline{X} - \text{cycles to failure at 5\% probability}}{\text{standard deviation}}$$

Therefore

$$\text{Cycles to failure at 5\% probability} = \overline{X} - (1.65) \,(\text{standard deviation})$$
$$= 5.4477 - (1.65)(0.0566)$$
$$= 5.3543 \qquad (\text{in terms of log scale})$$

The value is converted to true cycles to failure at 5 percent probability by taking the antilogarithm:

$$\text{antilog } 5.3543 = 226,100 \text{ cycles}$$

The cumulative probabilities for field and lab are drawn by plotting the model $F(i) = (i - 0.5)/n$ on the vertical axis and cycles to failure on the horizontal axis, as shown in Fig. 2.7. Other models, explained in Sec. 2.7, may also be used. If the two lines are parallel (or coincident), the correlation between the two tests is considered good. In this case, they are not quite parallel, indicating that the

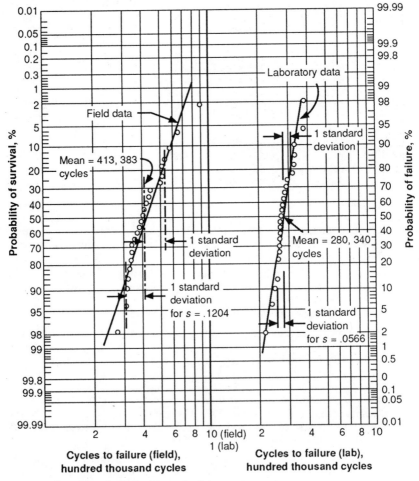

Figure 2.7 Log normal plot for Example 2.4.

field environment is loosely related to the lab test. But, since the failure modes are identical, the correlation can be considered acceptable (see Chap. 3 for a discussion of this point). Correlation can be quantified, if desired, by using the least squares method covered in standard books on statistics.

To establish the multiplication factor between the two plots, one can compare the means:

$$\text{Multiplication factor} = \frac{\text{mean of field plot}}{\text{mean of laboratory plot}}$$

$$= \frac{413{,}383 \text{ cycles}}{280{,}340 \text{ cycles}}$$

$$= 1.475$$

The log normal model seems valid since most of the points are on a straight line.

Note: If the two plots are not parallel, the validity of the multiplier becomes doubtful. See the discussion of accelerated testing in Chap. 3.

2.5 Exponential Distribution

The exponential distribution is the most widely used (and misused) distribution for reliability analysis. It applies when the failure rate is constant. Figure 2.8 shows some properties of this distribution. The failure rate is the fraction of failure per unit of time: say, percent per 1000 hours, per million hours, or per billion device hours. Other units, such as miles or cycles, can be used in place of hours. When the failure rate is defined as number of failures per billion device hours, it is expressed as FITS (failure units); 15 FITS means 15 failures per billion hours. Stewart Peck originated the FITS concept while at AT&T. That company now uses the concept widely, as do Mitsubishi[5] and others because the failure rates on many of their components are very low.

The failure rate λ is a parameter of the exponential distribution. The pdf for this distribution is given by the equation

$$f(t) = \lambda e^{-\lambda t}$$

where t is the time to failure. Failure rate for exponential distributions is often estimated as

$$\hat{\lambda} = \frac{\text{number of devices that failed}}{\text{total operating hours}}$$

Therefore, the probability of failure up to time t is

$$F(t) = \int_0^t \lambda\, e^{-\lambda t}$$

and the reliability is

$$R(t) = 1 - \int_0^t \lambda\, e^{-\lambda t}$$

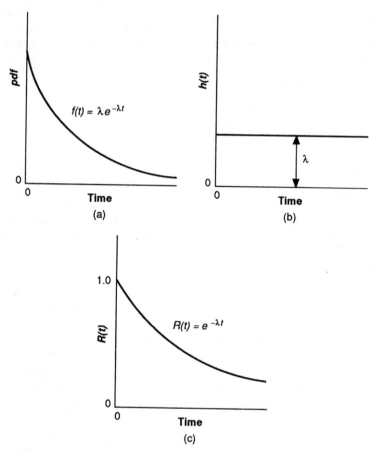

Figure 2.8 Some properties of the exponential distribution: (*a*) the probability density function; (*b*) the failure rate; (*c*) the reliability function.

In this equation λt is the average number expected to fail before time t. The only parameter of the exponential distribution is the failure rate λ.

Since λ is constant, most contracts use the reciprocal of failure rate, the mean time between failures (MTBF), as a measure of reliability. MTBF is estimated as

$$\text{Estimate of MTBF} = \hat{\theta} = \frac{\text{total operating hours}}{\text{number of devices failed}} = \frac{1}{\lambda} \quad (2.4)$$

The equation for reliability can be written in terms of MTBF as

$$\hat{R}(t) = e^{-t/\text{MTBF}} \quad (2.5)$$

Note: The term MTBF is used for repairable systems. For others, the appropriate term is mean time to failure (MTTF).

Example 2.5 Ten devices were tested. The test was stopped at the occurrence of the fourth failure. The hours to failures were 16, 40, 180, and 300. Six devices ran 300 hours without failure.

1. Calculate the MTBF.
2. If the sample MTBF is 584 hours, should the manufacturer guarantee repairs for failures up to 584 hours in the field?

solution The MTBF is estimated from Eq. (2.4):

$$\text{MTBF} = \frac{16 + 40 + 180 + 300 + 6 \times 300}{4} = \frac{2336}{4} = 584 \text{ hours}$$

(*Note*: This estimate is appropriate *only* for an exponential distribution.)

Suppose that the repairs are to be paid for by the manufacturer for the 584-hour duration. The estimated reliability is

$$\hat{R}(t) = e^{-t/\text{MTBF}} = e^{-584/584} = e^{-1} = .37$$

This shows that only an estimated 37 percent of the devices will survive 584 hours; 63 percent may fail during the warranty. Since most manufacturers will not be able to afford that many repairs, the manufacturer should not guarantee them for 584 hours.

Example 2.6 Suppose that the mean cycles between failures (equivalent to MTBF) of a product is 100,000 cycles. What should be the warranty period if the manufacturer wishes to pay for repair of no more than 10 percent of the devices?

solution Since $\hat{R}(t) = e^{-t/\text{MTBF}}$,

$$t = \text{MTBF} \times \ln\left(\frac{1}{R}\right)$$

where ln is the natural logarithm. In this example $R = 1 - .10 = .90$. Therefore,

$$t = 100,000 \ln\left(\frac{1}{.90}\right)$$

$$= 100,000 \times 0.1054$$

$$= 10,540 \text{ cycles}$$

Example 2.7 If a customer requires a guarantee during 100,000 cycles, what should be the MTBF goal for the supplier. The supplier is willing to pay for repairs of 5 percent of their product.

solution The reliability goal R is $1 - .05 = .95$. Since $R = e^{-t/\text{MTBF}}$,

$$\hat{\theta} = \frac{t}{\ln(1/R)}$$

$$= \frac{100,000}{0.0513} = 1,949,317 \text{ cycles}$$

Statistical estimation of the exponential parameter. The above calculations yield point estimates. If the limits are desired, then the following
calculations can be used.

For *failure-truncated testing*, in which the test is terminated at a
predetermined number of failures:

1. Two-sided limits for MTBF are

$$\frac{2T}{\chi^2_{\alpha/2,\,2r}} \leq \theta \leq \frac{2T}{\chi^2_{(1-\alpha/2),2r}} \tag{2.6}$$

where θ = true population MTBF
   α = probability that the true value of MTBF is outside the
        confidence limits
   T = total operating hours, cycles, or miles
   r = number of failures
   χ^2 = chi-square statistic for $2r$ degrees of freedom obtained
        from Table C in the appendix (see Ref. 2 for more de-
        tails)

2. The one-sided lower limit for MTBF is

$$\frac{2T}{\chi^2_{\alpha,2r}} \leq \hat{\theta} \tag{2.7}$$

For *time-censored testing*, in which the test is terminated at a predetermined time: [note the 2 $(r + 1)$ degrees of freedom in the lower bound]

1. Two-sided limits for MTBF are

$$\frac{2T}{\chi^2_{\alpha/2,2(r+1)}} \leq \hat{\theta} \leq \frac{2T}{\chi^2_{(1-\alpha/2),2r}} \tag{2.8}$$

2. The one-sided lower limit for MTBF is

$$\frac{2T}{\chi^2_{\alpha,2(r+1)}} \leq \hat{\theta} \tag{2.9}$$

Example 2.8 In Example 2.5, the test was failure-truncated; it was terminated
at the occurrence of the fourth failure with an MTBF estimate of 584 hours.
What are the 90 percent confidence limits for the true MTBF?

solution Equation (2.6) applies. Then the lower bound for $\hat{\theta}$ is

$$\frac{2T}{\chi^2_{.05,8}} = \frac{2 \times 2336}{15.507} = 301.28 \text{ hours}$$

and the upper bound on $\hat{\theta}$ is

$$\frac{2T}{\chi^2_{.95,8}} = \frac{2 \times 2336}{2.733} = 1709.48 \text{ hours}$$

Example 2.9 Fifty taxi drivers used a new component on automobiles to assess its MTBF. Together they accumulated 2,686,442 miles. At least 10 drivers drove more than 80,000 miles. There were no failures. The test was stopped at 93,000 miles. What is the minimum expected MTBF with 95 percent confidence?

solution The test is time- (miles-) censored, and Eq. (2.9) applies. Therefore, the one-sided lower limit for MTBF is

$$\frac{2 \times 2,686,442}{\chi^2_{.05,2(0+1)}} = \frac{5,372,884}{5.991} = 896,825 \text{ miles}$$

Note: The information about 10 drivers driving 80,000 miles was not used in the computation. This information is important for those who want to have an estimate of the component life. *MTBF* and *life* are not the same. See Chap. 3 for estimating life.

Example 2.10 What is the 1,000,000-mile reliability for the component in Example 2.9 if the customer is satisfied with a 70 percent confidence limit?

solution The $\hat{\theta}$ limit for 70 percent confidence is

$$\frac{2 \times 2,686,442}{\chi^2_{.30,2(0+1)}} = \frac{2 \times 2,686,442}{2.408} = 2,231,264 \text{ miles}$$

At the lower 70 percent confidence limit, the reliability is, from Eq. (2.5),

$$R(100,000) = e^{-100,000/2,231,264} = e^{-.0448} = .956$$

2.6 Weibull Distribution

There is a new trend toward use of the Weibull distribution. The distribution is not new; the theory was advanced in 1951, but applications have been mostly in the aerospace and automotive industries. The reason this distribution is getting wider acceptance is that it adequately fits many data. In addition, the exponential distribution is a special case of the Weibull distribution. Figure 2.9 shows some curves for the Weibull model. The commonly used Weibull model has two parameters:

β, *the shape parameter* The value of β determines the shape of a Weibull curve.

α, *the scale parameter*, also called the *characteristic life*. It is the life at the 63.2 percentile of the distribution. Curves of all shapes intersect at a common value of α when they are plotted as a cumulative distribution, as shown in Fig. 2.9*b*

Noteworthy properties of the Weibull distribution are

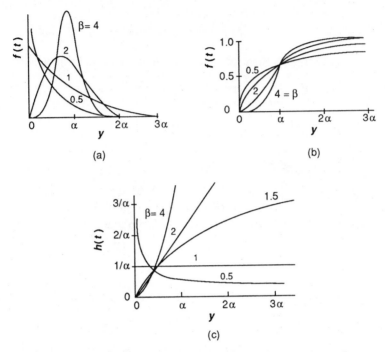

Figure 2.9 Weibull family of curves: (*a*) probability densities; (*b*) cumulative distributions; (*c*) hazard functions. (*Source: Wayne Nelson, How to Analyze Reliability Data, published by the American Society for Quality Control, 310 W. Wisconsin Ave., Milwaukee, WI 53203, 1983.*)

1. When the value of the shape parameter β is greater than 1, the rate of failure increases with time.

2. When β is less than 1, the failure rate decreases with time.

3. When β equals 1, the failure rate is constant, that is, the distribution is exponential.

Most Weibull distributions fitted to data have β values between 0.5 and 4. Values above 4 indicate a fast wearout, or the data may be corrupted.

The probability density function is

$$f(t) = \frac{\beta t^{\beta-1}}{\alpha^{\beta}} \exp\left[-\left(\frac{t}{\alpha}\right)^{\beta}\right] \qquad \text{for } t > 0 \qquad (2.10)$$

$$= 0 \qquad \text{otherwise}$$

where t is the time to failure.

Equation (2.10) is elaborate, but fortunately a simpler equation, that for the cumulative distribution function, is the one that is usually used to calculate reliability. The cdf is

$$F(t) = 1 - \exp\left[-\left(\frac{t}{\alpha}\right)^{\beta} \right]$$

Then reliability is

$$R(t) = \exp\left[-\left(\frac{t}{\alpha}\right)^{\beta} \right] \qquad (2.11)$$

This equation is revealing. Earlier it was said that the exponential distribution is a special case of the Weibull distribution. Substituting $\beta = 1$ into Eq. (2.11) reduces it to the traditional exponential equation

$$R(t) = \exp^{-}\left(\frac{t}{\alpha}\right)$$

Note that α is used instead of the usual θ.

Since the hazard rate is the ratio of $f(t)$ and $R(t)$, the Weibull hazard rate from Eqs. (2.10 and 2.11) can be written as

$$h(t) = \frac{\beta}{\alpha^{\beta}}(t)^{\beta-1} \qquad (2.12)$$

Example 2.11 Failure data for a component show $\beta = 1.5$ and characteristic life $\beta = 2000$ hours (a two-parameter Weibull distribution). Find:

1. The reliability at $t = 2000$ hours
2. The hazard rate at 500 hours
3. The hazard rate at 2000 hours

solution The reliability, from Eq. (2.11), is

$$R = e^{-(2000/2000)^{1.5}} = e^{-1.5} = .223$$

The 500-hour hazard rate, from Eq. (2.12), is

$$h(500) = \frac{1.5}{2000^{1.5}}(500)^{1.5-1.0} = \frac{1.5}{89,442.72} \times 22.361$$

$$= 0.000375 \text{ failure/hour}$$

The 2000-hour hazard rate is

$$h(2000) = \frac{1.5}{2000^{1.5}}(2000)^{1.5-1.0}$$

$$= \frac{1.5}{89,442.72} \times 44.721 = 0.00075 \text{ failure/hour}$$

Example 2.12   What values would be calculated in Example 2.11 if the failure density distribution was exponential?

solution   For an exponential distribution, $\beta = 1$. Then the reliability is

$$R = e^{-(2000/2000)^{1.0}} = e^{-1} = .37$$

Similarly,

$$h(500) = \frac{1.0}{2000^{1.0}} (500)^{1.0-1.0}$$

$$= \frac{1}{2000} \times 1 = 0.0005 \text{ failure/hour}$$

$$h(2000) = \frac{1}{2000^{1.0}} (2000)^0 = \frac{1}{2000} = 0.0005 \text{ failure/hour}$$

The values of $h(500)$ and $h(2000)$ are the same, since hazard rate is constant in an exponential distribution.

2.7   Data Analysis with Weibull Distribution

Mathematical estimation of the Weibull parameters α and β is cumbersome. Fortunately, a simple graphical method has been developed and is described here. Readers interested in mathematical estimation and other graphical methods such as hazard plotting can find more information in Refs. 3 and 6.

Weibull analysis applies chiefly to life test data from fatigue tests, although other models such as the log normal and gamma distributions may also apply. The Weibull distribution can be used for many kinds of reliability data when a single failure mode is being analyzed. There have been some misapplications in industry, and cautions will be indicated later.

Graphical analysis uses a special Weibull probability chart based on the Weibull mathematical model.[3,6] The chart in Fig. 2.10 is from an automotive company, but similar graph papers may be purchased.* The vertical axis is the cumulative probability plotting position $F(i)$. There are several models for $F(i)$. The most frequently used is the median rank model $F(i) = (i - 0.3)/(n + 0.4)$, where i is the order number when the failure times are arranged (ordered) from the lowest time to failure to the highest and n is the sample size.[7] Table D in the appendix contains computed values of the median rank for various sample sizes.

The x axis is a log scale for plotting time to failure (or cycles to failure or miles to failure). Each cdf shown in Fig. 2.9 plots as a straight

---

*A source is Technical and Engineering Aids for Management (TEAM), P.O. Box 25, Tamworth, New Hampshire. Sample charts are also found in J. R. King, *Probability Charts for Decision Making*, Industrial Press, New York, 1971.

Figure 2.10 Weibull probability graph.

line on the Weibull probability paper with a slope equal to its value of the Weibull parameter β. The life corresponding to 63.2 percent cumulative probability of failure is the value of α.

In graphically assessing reliability for a component or for a failure mode, one does not require the values of α and β. The $F(t)$ at any time t

can be obtained directly from the graph. The reliability is then found by a straightforward calculation: $R = 1 - F(t)$.

One of the advantages of Weibull analysis is that many times it does not require a large sample size. If a straight line is obtained with a sample size of 10, the chances are that the plot for 20 points will be very close to it. Of course, the plot with higher sample size is always more accurate. However, this luxury is not always available with engineering prototypes such as the experimental components of a large turbine for a power station. Sample sizes less than 10 have been used on prototype parts.

Plotting the data.   The plotting procedure below is used when a complete sample has failed. (For a partially failed sample, the procedure for censored data should be used, as described in Ref. 3.) An example will follow.

1. Put the times to failure in order, from lowest to highest. Assign order number to each failure sequentially.

2. Look up the median rank value from Table D in the appendix for each order number. The median rank is a plotting position on the y axis for $F(i)$. (Other models may be used such as those in Refs. 3 and 8.)

3. Plot all the points according to median rank value and time to failure.

4. Draw the line of best fit through the plotted points by using a transparent straight edge so that a roughly equal number of points is on each side of the line. The least square method may be used for more accuracy.

5. For a given time t on the x axis, find the value of $F(t)$ on the y axis.

6. Compute $\hat{R}(t) = 1 - F(t)$.

Example 2.13   A specification requires that no part should fail for 25 hours. Plot the hours-to-failure data for the part, tabulated below, on Weibull plotting paper. Using the graph, estimate:

1. The fraction expected to fail in 25 hours
2. The reliability if the customer wants failure-free operation for 25 hours

Specimen no.	Hours to failure
1	90
2	128
3	180
4	112
5	254
6	35
7	168
8	65
9	220
10	108

solution Order the data and assign the median rank from Table D in the appendix for sample size 10 as follows:

Hours to failure	Order no.	Median rank
35	1	6.7
65	2	16.2
90	3	25.9
108	4	35.5
112	5	45.2
128	6	54.8
168	7	64.5
180	8	74.1
220	9	83.8
254	10	93.3

Plot the graph using hours to failure on the x axis and median rank on the y axis. Figure 2.11 shows the resulting Weibull plot.

The fraction expected to fail in 25 hours, from the graph, is 2.3 percent (point A on the y axis). The reliability for 25 failure-free hours = $1 - .023 = .977$.

Cautions: Weibull analysis is best suited for fatigue tests. Therefore, it is best suited for component testing. At system level, the failure may be not from fatigue but from infant mortality problems of soldered and welded joints (Chap. 3). The Weibull model might be expected to suit the system failure data, but the analyst should judge cautiously. Moreover, at system level the nature of failures differs from time to time. The model may therefore be an acceptable indicator of past performance, but its validity as a predictor of later life is questionable.

Figure 2.11 Weibull plot for Example 2.13.

2.8   Discrete Distributions

Most reliability work is based on continuous distributions. Sometimes, however, the variables have certain discrete values but cannot have the values in between. For example, the number of failures of a system cannot be a fraction. Some qualitative features also take certain discrete values. For example, a product may have only two reliability levels—acceptable and unacceptable. These two states can be assigned values of 1 and 0, but is not meaningful to assign values between these two numbers. This chapter covers two such discrete distributions—the binomial and Poisson distributions—because they are widely used. References 1 and 3 cover these and other distributions in more detail.

2.8.1   Binomial distribution

The binomial distribution applies to situations where there are only two mutually exclusive outcomes. For example, a product is reliable or not reliable; a product meets the specification or does not meet the specification; blood samples are positive or negative; a malfunction is caused by high pressure or low pressure. The term *mutually exclusive* means the two outcomes cannot occur simultaneously.

The binomial distribution can be applied to modeling of the reliability (or the probability of failure) of fault-tolerant systems. These systems usually contain hardware or software redundancy. If there are five computers on a space shuttle and if at least three must work during a mission, the binomial distribution can be used to compute the system reliability. The general expression for reliability of a system with exactly x components out of N working is

$$R(s) = \sum \binom{N}{x} R^x (1 - R)^{N-x} \qquad (2.13)$$

where R is the component reliability and N represents the expression $N! \, [(x!)(N - x)!]$.

Example 2.14   A computer has five processors. At least three must be in working condition during a mission. If each processor has an estimated reliability of .9, what are the chances of the system success?

solution   The system will work if three processors survive or four processors survive or all five survive. Thus, from Eq. (2.13),

$$R(s) = \binom{5}{3}(.9)^3(.1)^{5-3} + \binom{5}{4}(.9)^4(.1)^{5-4} + \left(\frac{5!}{5!0!}\right)(.9)^5(.1)^{5-5}$$

$$= \frac{5!}{3!2!}(.9)^3(.1)^2 + \frac{5!}{4!1!}(.9)^4(.1)^1 + \left(\frac{5!}{5!0!}\right)(.9)^5(.1)^0$$

$$= (10)(.729)(.01) + (5)(.6561)(.1) + (1)(.5905)(1)$$

$$= .9915$$

Example 2.15 The space shuttle has three surviving processors. If the outputs of at least two processors must match, what is the system reliability if each processor is .99 reliable?

solution $N = 3$, since only three processors are available. But since the outputs of only two processors are used, we consider only $x = 3$ and $x = 2$ in calculating system reliability from Eq. (2.13):

$$R(s) = \binom{3}{2}(.99)^2(.01)^{3-2} + \binom{3}{3}(.99)^3(.01)^{3-3}$$

$$= \frac{3!}{2!1!}(.99)^2(.01) + \frac{3!}{3!0!}(.99)^3(.01)^0$$

$$= (3)(.9801)(.01) + (1)(.9703)(1)$$

$$= .9997$$

2.8.2 Poisson distribution

The Poisson distribution is used for estimating the number of random failures or occurrences in a given time when their likelihood is small. It should not be used for predicting fatigue-related failures or premature failures from manufacturing faults. It is applied when the sample size is at least 20 and the probability of occurrence is less than .05. The population size can be very large—as much as infinity. Under these conditions, it approximates the binomial distribution.

The Poisson distribution assumes that the probability of occurrence is exponential, but while the exponential distribution is continuous, the Poisson is a discrete distribution. The general expression for x occurrences in a given time is

$$f(x) = \frac{a^x e^{-a}}{x!} \tag{2.14}$$

where a is the expected number of occurrences based on historical data and x is the number of occurrences of a failure.

Example 2.16 Historical records show the following information on unscheduled outages of a power plant:

Number of out-ages in a week	Frequency
0	2
1	9
2	4
3	2
4	1

1. What is the chance of no outage?
2. What is the probability of two or less outages?

solution   First find a, the expected number of outages per week:

$$\text{Total number of outages} \over \text{Total number of weeks}$$

$$a = \frac{2 \times 0 + 9 \times 1 + 4 \times 2 + 2 \times 3 + 1 \times 4}{18} = 1.5$$

The probability of two or less occurrences is then calculated from Eq. (2.14) as the sum of the probabilities of zero, one, and two failures:

$$f(2 \text{ or less}) = \frac{(1.5)^0 \, e^{-1.5}}{0!} + \frac{(1.5)^1 \, e^{-1.5}}{1!} + \frac{(1.5)^2 \, e^{-1.5}}{2!}$$

$$= e^{-1.5} (1 + 1.5 + 1.125)$$

$$= .223 \, (3.625) = .8084$$

The chance of no outage is the first term of the sum above:

$$f(0) = \frac{(1.5)^0 \, e^{-1.5}}{0!} = .3345$$

2.9   Topics for Student Projects and Theses

1   Explore why textbooks emphasize the exponential distribution in reliability models. Is it a correct approach? (Read Chap. 3 for more information.)

2   Describe at least four statistical models not covered in this chapter.

3   All quality control work is based on the assumption that a variable is distributed normally (or skewed). Why are the exponential, log normal, and Weibull distributions not considered? (See Chap. 6.)

4   Develop applications of the Poisson distribution for early design work.

References

1. G. J. Hahn and S. S. Shapiro, *Statistical Models in Engineering*, John Wiley and Sons, New York, 1967.
2. Harrison M. Wadsworth, *Handbook of Statistical Methods for Scientists and Engineers*, McGraw-Hill, New York, 1990.
3. Wayne Nelson, *Applied Life Data Analysis*, John Wiley and Sons, New York, 1982.
4. S. S. Shapiro, "How to Test Normality and Other Distributional Assumptions," American Society for Quality Control, Milwaukee, 1980.
5. Hiroshi Matsumura, Nobuhisa Maki, and I. Matsumoto, "Strategic Performance of Reliability Evaluation for New Sophisticated VLSI Product," *Proceedings of Inter-*

national Conference on Quality Control, Union of Japanese Scientists and Engineers, Tokyo, 1987, pp. 649–650.

6. Walodi Weibull, "A Statistical Distribution Function of Wide Applicability," *Journal of Applied Mechanics*, September 1951, pp. 293–297.
7. Charles Lipson and Narendra Sheth, *Statistical Design and Analysis of Engineering Experiments*, McGraw-Hill, New York, 1973.
8. L. G. Johnson, *The Statistical Treatment of Fatigue Experiments*, Elsevier, New York, 1964.
9. M. O. Locks, *Reliability, Maintainability, and Availability Assessment*, Hayden, Rochelle Park, N.J., 1973.

Further Reading

Dixon, Wilfrid J., and Frank J. Massey, Jr., *Introduction to Statistical Analysis*, 4th ed., McGraw-Hill, New York, 1983.

Gross, Alan J., and Virginia A. Clark, *Survival Distributions: Reliability Applications in the Biomedical Sciences*, John Wiley and Sons, New York, 1975.

Kapur, K. C., and L. R. Lamberson, *Reliability in Engineering Design*, John Wiley and Sons, New York, 1977.

Lee, Elisa T., *Statistical Methods for Survival Data Analysis*, Lifetime Learning Publications, Wadsworth Inc., 1980.

Lewis, Elmer E., *Introduction to Reliability Engineering*, John Wiley and Sons, New York, 1987.

Nelson, Wayne, *Accelerated Life Testing*, John Wiley and Sons, New York, 1990.

O'Connor, Patrick D. T., *Practical Reliability Engineering*, 2d ed., John Wiley and Sons, New York, 1985.

Reliability Engineering

3.1 Reliability Principles

Lord Kelvin once said, "I often say that when you can measure what you are talking about and express it in numbers, you know something about it." This is certainly true of product design. An engineer may make a product work in the laboratory, but the real criterion is how many of the units produced work in the field. If less than half the units work, as in the case of the Adam computer,[1] the engineering department can hardly expect accolades.

Even though some prototypes may work well in the laboratory, others may have seemingly minor problems. Such problems are often ignored and attributed to human error. Similar errors and new ones can occur on the manufacturing floor. They compound, resulting in very low reliability and eventual loss of market share; they can reduce reliability to a very low number.

Therefore, the proportion of successful products should be documented in design, in manufacturing, and in the field. The documentation reveals which life-cycle phase is responsible for the downfall.

Reliability engineering in the literature is mostly perceived as a statistical tool. The truth is that statistics is mostly used for measuring, analyzing, and estimating various parameters. Most of the true reliability work is done during the development of a design, long before test data are available for statistical manipulation. True reliability is that built into the design and is called *inherent reliability*. It is the highest reliability a product will have. Neither responsive manufacturing nor preventive maintenance can raise it above this level.

As the product passes through the production stage, its inherent reliability tends to be degraded, although good screening and process control programs can reduce or prevent degradation. Reliability is further degraded in the field because of unanticipated stresses or environments, inadequate procedures, and poor maintenance. If the product's inherent reliability is extremely high, it can sometimes allow for field degradation and still meet the customer's expectation.

Reliability is defined as the probability that a product will perform its intended function for a given time period under a given set of conditions. The definition assumes the product is in working condition at the outset. Time in this definition refers to such durations as mission time, warranty period, a predefined number of cycles, and even a complete lifetime. A product can also have several missions. In that case, a separate set of requirements may be developed for each mission.

The poorly understood bathtub curve.   To understand reliability, it is necessary to understand the concept of the so-called bathtub curve. This is a plot of hazard rate versus life (Fig. 3.1). It consists of three regions, each having different hazard rates. In combination, the distributions form a single smooth curve that resembles a bathtub profile. Reliability is a function of hazard rate in each region.

The failures in region I are premature and are likely to be present in the field, unless due care is taken in manufacturing. They are usually a result of shoddy manufacturing, poor workmanship, and excessive variation in the material. They can be overcome by process control and burn-in and screening programs. Failures in this region are called *infant mortality failures.*

Region II, called the *useful design life* period, represents the real

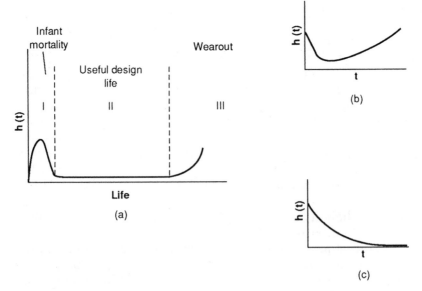

Figure 3.1 Typical failure rates versus life bathtub profiles. (*a*) Electronic equipment; (*b*) mechanical equipment; (*c*) software.

failure rate the customer can expect. There is a general misunderstanding in industry about this region. For a good design, this region has a very low hazard rate and the curve is approximately flat (Fig. 3.1a), indicating a constant failure rate. The misunderstanding is that many practitioners assume that the region always has a constant hazard rate even though many times it does not (Fig. 3.1b and c). To make matters worse, some practitioners assume a constant hazard rate even in region I, where such a condition is a rarity. No wonder that failure predictions made in industry rarely correlate with field results. To make meaningful predictions, a reliability analyst must consider the impact of infant mortality failures (usually from production causes). Because production problems are different at various times, the distribution of infant mortality failures is unstable. If region I failures are allowed to go to the customers, reliability prediction is optimistic.

To simplify computations, the industry uses *mean time between failures* (MTBF), the reciprocal of hazard rate, as a measure of reliability. Many misunderstand this term (including many writers of government documents) as an equivalent of life. Actually, MTBF has absolutely no relation to life. MTBF is only a function of the gap between the x axis and the bottom of the bathtub. This gap is independent of the life or the length of the bathtub! This gap (the hazard rate) is usually a function of design.

Region III is the *wearout* phase. Its shape is a function of design. At system level, the onset of wearout can be pushed to the right if the components are replaced before they fail. In general, preventive maintenance can extend the useful life of a system.

Reliability improvement techniques.   There are many design improvement techniques for reliability. The widely used techniques are

- *Zero failure design*, where causes of critical failures are entirely eliminated by design. (The product can fail from unanticipated causes.) A similar concept in the aircraft industry is called *safe life design*, where the components are qualified for twice the expected life. This allows a failure to occur but not during the useful life.

- *Fault tolerance*, where redundant elements are used to switch over to a backup or alternative mode.

- *Derating*, where a component is used much below its capability rating.

- *Durability*, where a component is designed to have longer useful life or is designed for damage tolerance (such as aircraft structures).

- *Design with product safety margins* for all applicable worst-case stresses and environments.

Major analytical tools needed to achieve these objectives are

Design reviews

Reliability allocation, prediction, and modeling

Failure-mode, effects, and criticality analysis

Fault-tree analysis (see Chap. 5)

Sneak circuit analysis

Worst-case analysis

Statistical analysis of failure distributions

Quality function deployment (see Chap. 6)

Robust design methodology using design of experiments (see Chap. 6)

Most of these techniques are covered in this chapter; others are covered in the chapters indicated above. "References" and "Further Reading" at the end of the chapter give sources for more information.

3.2 Reliability in the Design Phase

The author's experience shows that as much as 60 percent of failures can be prevented by making changes in the design. In all design improvements, the designer can choose a strategy for preventing failure according to the following order of precedence:

1. Improve the design to eliminate the failure mode.

2. Design for fault tolerance.

3. Design for fail-safety.

4. Provide early warnings of failure through fault diagnosis.

If these strategies are not viable, the designer may choose reliability-centered preventive maintenance (see Chap. 7) and issue special maintenance instructions to the user. These instructions should be presented in such a way that users will be sure to read and understand, and follow them.

The activities in the design stage start with specifying the reliability requirements. These specifications are implemented in design through reliability analysis and are verified by testing.

3.2.1 Writing reliability specifications

Reliability requirements should be an integral part of engineering specifications. As the definition of reliability implies, at least four features must be identified:

1. The probability of successful performance of the product

2. The functions (missions) to be performed

3. The intended time (such as mission time)

4. Operating conditions

These requirements give a skeleton structure for a reliability specification. But a design engineer needs ideas to meet the specification requirements.

Consider a reliability specification for a robot for welding automobile chassis. The probability of successful performance can be specified in several ways. It may be specified as *availability*, the probability that the robot is in working condition during the scheduled hours only. It may be specified as several different probabilities for various missions; some functions may be less important than others. Limits on applications may be specified; for example, the robot may not be able to weld certain steels.

The intended time can be specified as hours of actual operations, number of cyclical loads, or calendar time. The operating environment should be specified so that qualification tests can be developed. In this case, the environment would include dust, chemicals, sparks, humidity, vibration, shock, and perhaps an unskilled machine operator. The specification should also include software reliability. Chapter 9 contains a section on this aspect.

Example 3.1   An automotive cranking system consists of engine, starter motor, and battery. The reliability requirement for the system in an engineering performance standard reads: "There shall be a 90 percent probability [probability of success] that the cranking speed is more than 85 r/min after 10 seconds of cranking [mission] at − 20°F [environment] for a period of 10 years or 100,000 miles [time]. The reliability shall be demonstrated at 95 percent confidence."

3.2.2 Conducting design reviews

Design reviews are the most important part of product engineering. They include discussions of the results from various analyses, informal brainstorming, and review of checklists and project controls.

Design reviews should be a team effort headed by someone who can provide unbiased leadership. With proper leadership, design productivity will increase significantly.

The purpose of design reviews is to challenge the design from different viewpoints. This can happen only when designers and nondesigners are encouraged to think creatively and to advance fresh views. The biggest benefit of design reviews is that engineering changes can be made early, when they will have the greatest effect in reducing life-cycle costs. Darch[2] presents several noteworthy viewpoints on design reviews:

- Design reviews provide a forum for discussion of interface problems and their resolution.
- Design reviews provide a common base for future work.
- As design proceeds toward completion, design reviews continue to reduce life-cycle costs.
- The nature of the design review process is such that, after a stage is completed, the results from that stage are accepted as final and the next stage proceeds from there.

Decisions resulting from design reviews at any stage should be given serious consideration at the top management level. A design review is not merely a review of product performance, but a risk assessment process with major effects on future costs. The decisions will determine the inventory of spares, skill levels of operating and maintenance personnel, test equipment required in the field, and many support requirements. Therefore, before design review decisions are approved, the life-cycle costs they will incur should be estimated.

Preliminary design review.   A preliminary design review should be held to select strategies for developing design improvements. The following strategies are available:

1. Modular design
2. Fault-tolerant design
3. Fault isolation
4. Fault masking
5. Built-in testing
6. Self-checking
7. Self-monitoring
8. Derating
9. Self-healing
10. Design for ease of maintenance
11. Design for ease of inspection
12. Design for throwaway instead of repair (for example, light bulbs)
13. Testable design
14. Maintenance-free design
15. Zero-failure design
16. Built-in retry (software)
17. Design for damage detection

18. Top-down structure (software)

Lessons learned and checklists. It is always a good idea to start a design review with two important items, (1) the lessons learned since the previous review and (2) a checklist subject to ongoing development.

The lessons-learned discussion includes examinations of failure histories, competitive analyses, and interviews with dealers, distributors, and users of similar products. It is important to eliminate from the discussion, as far as possible, failure modes common to the designs of all competitors. At the same time, the strengths of the competitors should be discussed. It may be possible to introduce new strengths into a product, such as maintenance-free, fault-tolerant, or modular design.

Checklists are powerful tools in getting a review done faster, but they do not cover all design concerns. Additional brainstorming is required.

The following generic checklist may be useful as a start.

1. Are reliability, maintainability, availability, and safety features included in the specification?
2. Is failure-mode, effects, and criticality analysis (FMEA) planned in early design?
3. Can fault-tree analysis be used to improve the design?
4. Are component interchangeabilities analyzed?
5. Are safety margins adequate?
6. Is software reliability specified?
7. Is testability analysis required?
8. Is fault isolation capability needed?
9. Do electronic circuits have adequate clearances between them?
10. Are software logic concerns independently reviewed?
11. Has software coding been thoroughly reviewed?
12. Are qualification test plans done well in advance?
13. Are production tests planned and reviewed?
14. Are redundancies considered for software?
15. Are redundancies considered for hardware?
16. Has the design been critiqued for human mistakes?
17. Are designers familiar with the human engineering guidelines contained in Mil-Hdbk-1472?

18. Is reliability-centered maintenance considered in design?

19. Is competitive analysis for reliability done?

20. Are there long-term problems with design such as dendritic growth on integrated circuits?

21. Is the product analyzed for corrosion failures?

22. Is there a statistical model for analyzing test data?

23. Are the switches for backup devices reliable? Do they need maintenance?

24. Are protective devices such as fuses, sprinklers, and circuit breakers reliable? Do they need reliability-centered maintenance?

25. Is the product performance evaluated against radiation and accidental damages?

26. Does the product need to withstand earthquakes and unusual loads?

27. Are component tolerances optimized to give the desired reliability?

28. Has the product been tested for environments of the countries to which it will be exported?

29. Can manufacturing personnel introduce any defects? Can they be prevented by design?

30. Can maintenance personnel introduce defects? Can the product be designed to prevent this possibility?

31. Can the operator introduce wrong inputs? If so, can the product be designed to switch to a fail-safe mode?

32. Can a single component cause the failure of a critical function?

33. Is crack growth and damage tolerance analysis required?

34. Do components need corrosion protection?

35. Can applied loads be larger in the future (such as heavier trucks on a bridge)?

36. Are there unusual environments not already considered?

37. Are inspection provisions made for detecting cracks, damage, and flaws (for aircraft)?

38. Does design require crack-retardation provisions?

3.2.3 Reliability allocation

The process of allocation establishes a hierarchy of design requirements. The purpose is to distribute operational reliability goals to subsystems, subassemblies, and all the way down to components. Compa-

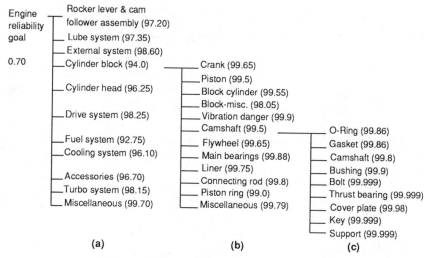

Figure 3.2 Allocation by reliability goals.

nies that do not allocate reliability rarely achieve the desired overall reliability. For example, a company making electronic warfare equipment struggled to get an MTBF of 360 hours but rarely achieved more than 40 hours MTBF. It always complained about suppliers' poor performance. Yet, on investigation, it was found that suppliers were never told their MTBF goals.

Allocations start with the system goal. For example, as shown in Fig. 3.2, designers of an engine with 70 percent reliability over 10 years will have to allocate higher reliabilities among subsystems, since the product of subsystem reliabilities is equal to the system reliability. Similarly, each subsystem reliability should be allocated among its subassemblies (only one breakdown is shown in the figure). The allocation continues until all the components are accounted for. Note that reliabilities as high as 99.999 are required to get only 70 percent reliability at engine level.

Where subsystems and components have constant failure rates, the allocation can be in terms of failure rate. This is more convenient than using reliabilities because failure rates for subsystems can be added instead of multiplied. For a serial system, the MTBF is first converted to a failure rate, which is allocated among the subsystems and components. At the completion of the analysis, the failure rates are converted back to MTBF. Figure 3.3 shows such an allocation for a serial system.

3.2.4 Reliability modeling

Reliability modeling is done in early design to assess alternative product reliability approaches. There are two common models:

Figure 3.3 Allocation by failure rates.[3]

1. The basic reliability model, popularly called the *series model*
2. The fault-tolerance model, popularly called the *parallel model*, in which some or all portions may be redundant

Series model. If a product will fail when any of its components fails, it is called a series model. For example, if any component of a bicycle fails, then the entire bicycle fails. Such a model can be represented by a chain. If any link in the chain fails, the entire chain has failed. Figure 3.4 shows series as well as several parallel models.

In the series model, since all the components must work, the system reliability is a function of all the reliabilities. The system reliability is represented by the equation

$$R(s) = R_a \cdot R_b \cdot R_c \cdots R_n \tag{3.1}$$

Figure 3.4 Series system (*a*) and parallel systems (*b,c,d*).

If there are four components with reliabilities .90, .80, .90, .70, respectively, the resultant reliability is

$$R(s) = .90 \times .80 \times .90 \times .70$$

$$= .4536$$

This computation shows that the more components there are of a given type, the worse the system reliability will be. Therefore, a design goal is to minimize the number of components for a series system. For a parallel system this may not be the case.

Parallel system. In a series system, the reliability degrades rapidly with increase in number of components; each fraction multiplied by another fraction results in a smaller fraction. To increase reliability one can back up a component of a series (Fig. 3.4b), or back up the entire chain with another chain (Fig. 3.4c), or individually back up each component (Fig. 3.4d). The reliability of a parallel system is computed from the relation

Probability of system working + probability of not working = 1

For Fig. 3.4b, the only time the system will not work is when both components have failed. The probability of the entire system not working is $(1 - R_a)(1 - R_b)$. From the above relation, the probability of the system working is

$$R(s) = 1 - (1 - R_a)(1 - R_b)$$

In general,

$$R(s) = 1 - (1 - R_a)(1 - R_b)(1 - R_c) \cdots (1 - R_n) \qquad (3.2)$$

From Eq. (3.2), the calculations for parallel models are Fig. 3.4b:

$$R(s) = 1 - (1 - .9)(1 - .9) = .99$$

Fig. 3.4c:

$$R(s) = 1 - (1 - R_a \cdot R_b \cdot R_c \cdot R_d)(1 - R_a \cdot R_b \cdot R_c \cdot R_d)$$

$$= 1 - (1 - .9 \times .8 \times .9 \times .7)(1 - .9 \times .8 \times .9 \times .7)$$

Fig. 3.4d:

$$R(s) = [1 - (1 - R_a)(1 - R_a)][1 - (1 - R_b)(1 - R_b)]$$

$$\cdot [1 - (1 - R_c)(1 - R_c)][1 - (1 - R_d)(1 - R_d)]$$

$$R(s) = [1 - (1 - .9)(1 - .9)][1 - (1 - .8)(1 - .8)]$$
$$[1 - (1 - .9)(1 - .9)[1 - (1 - .7)(1 - .7)]$$
$$= .99 \times .96 \times .99 \times .91$$
$$= .8562$$

3.2.5   Reliability prediction

Reliability modeling establishes how a system is configured. Reliability predictions estimate the reliability of the configuration to assure it will satisfy the allocated operational goals. Most practitioners of electronic equipment design use Mil-Hdbk-217,[4] based on an exponential failure distribution, to predict reliability. They assume the system is a series system and usually overlook redundant components. It is amazing that some experienced engineers, not just beginners, do this. Perhaps no harm is done because this practice gives the worst reliability estimate.

One reason this practice is so well accepted, even when the series assumption is wrong, is that it makes the reliability calculation so simple. The reliabilities can be simply multiplied to estimate system reliability, as in Eq. (3.1). If the components are also assumed to have an exponential failure distribution, the model becomes dramatically simple:

$$R(s) = R_a \cdot R_b \cdot R_c \cdots R_n = e^{-\lambda_a t} \cdot e^{-\lambda_b t} \cdot e^{-\lambda_c t} \cdots e^{-\lambda_n t}$$
$$= e^{-(\lambda_a + \lambda_b + \lambda_c + \cdots + \lambda_n)t}$$

This model shows that the failure rates can simply be added to give a total failure rate. The final model is

$$R(s) = e^{-(\Sigma \lambda)t}$$

The truth is that a good product should have a very low constant (or decreasing) failure rate, and therefore this model should be the goal. If (1) the infant mortality failures are completely screened and (2) the wearout process does not accelerate until after useful life is experienced, this model will apply, even for mechanical components. Most practitioners do not verify the validity of this model. They assume that, if there are hardly any failures during the useful life, the failure rate can be regarded as a very small constant.

Caution: Estimating reliability by the procedure above is only for the hardware. Field reliability is also degraded by manufacturing errors, maintenance errors, improper use, and software-induced errors.

In estimating failure rates from Mil-Hdbk-217, reasonable care should be exercised. The failure rate data can be 8 to 10 years old, the data are collected under varied conditions, and factors such as transient line voltages and electromagnetic interference are not considered. The best way to estimate failure rates is to build an in-house data bank using models similar to those in Mil-Hdbk-217. Also, before the estimates are made, one should consider the fact that the hazard rates for some electronic components are not constant. A military specification on capacitors[5] claims the failure rate for such devices has never been constant. It has always been a decreasing function of time. These considerations suggest that the predictions from Mil-Hdbk-217 are not reliable. The reader also should recognize that field reliability is influenced by human error, software, preventive maintenance, and the operating procedures. These features are so qualitative that it is impractical to predict field reliability.

One big advantage of failure rate models is that they make designers aware of some factors influencing the failures and their impact on the product. Figure 3.5 gives failure rate models for an electronic and a mechanical component. The electronic component model is from Mil-Hdbk-217. The model for the mechanical component is from a draft copy of a handbook for mechanical components being developed under contract to the U.S. Navy and U.S. Army.[6]

Example 3.2   A system for the U.S. Air Force consisted of 10 circuit boards to monitor 10 facilities independently (Fig. 3.6a). Each circuit board had about 150 components in the power supply region, about 100 components in the input-output region, and 70 components in the communication portion.

Since each facility had a separate circuit board assigned to it, troubleshooting was easy. The maintenance people knew right away which circuit board had failed. The reliability model for this system is a series model. Prediction from Mil-Hdbk-217 showed the circuit board reliability over a 5-year period to be .83. This gave a system reliability R = .155.

It was later decided that the cause of the low system reliability (the chance of failure was 15.5 percent) was the presence of too many components in series. A design engineer proposed that, since the power supply requirement is same for all the circuit boards, why not have one common external power supply. But if the single power supply failed, all the circuit boards would become nonfunctional. To overcome this problem, a redundant power supply was added. The new configuration is shown in Fig. 3.6b. The reliability is now .68 versus .155 before. This is a good example for those who do not believe that reliability reduces cost; in the improved version there are 2000 components versus 3200 in the original design. In addition, there is a 339 percent gain in system reliability. (*Note:* The maintenance cost may go up slightly to assure the redundant power supply is always in working condition).

3.2.6   Failure-mode, effects, and criticality analysis

When all obvious design concerns are seen to, it is time to run a detailed analysis (Fig. 3.7). Failure-mode, effects, and criticality analysis (FMECA) is a technique for microlevel analysis of the design. It starts

Part operating failure rate model (λ_p):

$$\lambda_p = \pi_Q \, (C_1 \pi_T \pi_V + C_2 \pi_E) \, \pi_L \text{ failures/10}^6 \text{hours}$$

where

λ_p is the device failure rate in F/10^6 hours
π_Q is the quality factor
π_T is the temperature acceleration factor, based on technology
π_V is the voltage stress derating factor
π_E is the application environment factor
C_1 is the circuit complexity factor based on transistor count as follows:
    for 1 to 100 transistors, $C_1 = 0.01$
    for > 100 to 300 transistors, $C_1 = 0.02$
    for > 300 to 1000 transistors, $C_1 = 0.04$
C_2 is the package complexity failure rate
π_L is the device learning factor

(a)

$$\lambda_{BE} - \lambda_{BE,B} \left(\frac{L_A}{L_S}\right)^{\gamma} \left(\frac{A_E}{.006}\right)^{2.36} \left(\frac{v_o}{v_L}\right)^{0.54} \left(\frac{C_{AL}}{60}\right)^{2/3} \left(\frac{M_O}{M_F}\right) \cdot C_{CW}$$

where λ_{BE} – predicted failure rate of bearing, using actual conditions, per million hours of operation
       γ – 3.33 for roller bearing; 3.0 for ball bearings
       L_A – actual radial load, lb
       L_S – specification radial load, lb
       A_E – alignment error, rad
       v_o – specification lubricant viscosity, lb-min/in^2
       v_L – operating lubricant viscosity, lb-min/in^2
       M_O – material factor, base material, lb/in^2, yield strength
       M_F – material factor, operating material, lb/in^2, yield strength
      C_{AL} – actual contamination level, micrograms/meter^3
      C_{CW} – water contamination factor

(b)

Figure 3.5 Failure rate models. (*a*) Electronic component; (*b*) mechanical component.

at the component level to determine what could go wrong with each component and its effects on the system. Unfortunately, many companies spend a substantial amount of money on FMECA (which is a good idea) but do not know what to do with the information (a gross misuse of this technique). An engineer showed the author an FMECA of a circuit board having over 4000 failure modes. When asked what he did

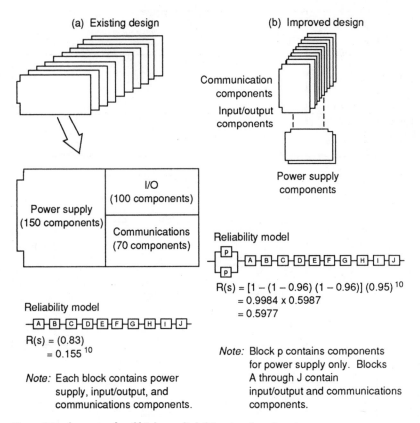

Figure 3.6 An example of higher reliability at reduced cost.

with the information, the engineer replied, "Nobody asked me to do anything with it." It was another nice-looking report that, after a 3-month effort, never saw daylight. Such a response is fairly common. Some use FMECA to predict reliability only. The right way to use FMECA is to improve the product design with this knowledge. Prediction should be treated as a by-product.

The original FMECA technique, suggested in Mil-Std-1629,[7] is to systematically analyze each component by using a format such as that in Fig. 3.8. The number and headings of the columns will vary with the application.

Further improvements in FMECA methodology are proposed here. For example, Mil-Std-1629 suggests identifying failure modes related to hardware failures only, while this author recommends treating human factors and workmanship problems, such as poor solder joints and welded joints, as a part of FMECA. This can be accomplished by dividing the analysis into two units: one for the product design, called

Criticality Analysis

System _____
Indenture level _____
Reference drawing _____
Mission _____

Date _____
Sheet _____ of _____
Compiled by _____
Approved by _____

Identification number	Item/functional identification (nomenclature)	Function	Failure modes and causes	Mission phase/ operational mode	Severity class- ification	Failure probability / Failure rate data source	Failure effect probability (β)	Failure mode ratio (α)	Failure rate (λ_p)	Operating time (t)	Failure mode criticality no. $C_m = \beta\alpha\lambda_p t$	Item criticality no. $C_? = \Sigma(C_m)$	Remarks

Figure 3.7

① Item part no.	② Block diag. ref.	③ Functions	④ Failure modes	⑤ Failure mechanism (Cause)	⑥ Effect on			⑦ Interface effects	⑧ Criticality			⑨ Corrective action		⑩ Item close sign off or rationale for other action
					Local	Higher levels	End item		Severity	Freq.	Detect-ability	Time avail. for corrective action	Recommendation	

Figure 3.8   FMECA matrix.

design FMECA and the other for manufacturing, called *process FMECA*. This chapter covers design FMECA. Process FMECA is covered in Chap. 6.

The documentation in forms like that in Fig. 3.8 is essential because it can be used for several other analyses such as safety, maintainability, quality, and logistic support. The procedure for filling in the form follows.

Column 1: Item part number. Write the component identification number. It is obtained from the bill of materials or indented parts list. Make sure *all* components are listed, including bolts, nuts, washers, wires, and screws; the purpose of the form is to help engineers analyze *systematically*. A part such as a nut may seem trivial, but improper clearance with a bolt might result in a catastrophic accident. (Several engines in U.S. Air Force jets fell off because of improper clearance. In fact, in 1987, 60 percent of the Air Force's engine inventory was found suspect of such problems.) Most defense contractors perform analysis only on components they think are important at a functional level, but they risk overlooking some critical components. This latter approach is also acceptable in Mil-Std-1629, especially for systems with components in thousands.

Column 2: Block diagram reference. This column is for identifying where the component exists in the block diagram, logic diagram, or assembly drawing. This column is optional.

Column 3: Function. The part function is described to give the analyst the clues to failure modes. Any defect or action which prevents the function from being performed is a failure mode. For example, if the function of a transistor is to act as a switch, then insufficient current, in addition to opens and shorts, constitutes a failure mode.

Column 4: Failure modes. A part will have several failure modes that can impede its function. The simplest example is a bolt whose function is to hold a cover plate in a vibration environment, in combination with a nut. Failure modes such as wrong pitch diameter, too much clearance between it and the nut, and improper plating can impede its function. Robust design considers not only how the part can fail or what can go wrong with it, but also the effects of human operators such as an assembler who forgets to assemble the bolt altogether.

Describing failure modes is often difficult and complex. They can be different in various mission phases. For example, a spacecraft can have one failure mode on the launching pad, another in geosynchronous orbit, and yet another in the reentry phase. Such was the case with a very tough and lightweight Kapton-insulated wire developed for Spacelab. The wire had excellent endurance properties, but it burst like fireworks in the zero gravity of space in the presence of moisture or any liquid, according to a NASA internal memo in the author's files.

Failure-mode prediction can be difficult even in a single-mission-phase product. Figure 3.9a shows a pressurized joint on a solid booster rocket. It shows two seals. The secondary seal is supposed to protect the system if the primary seal fails. Unpressurized, the secondary seal does what it is supposed to do. But pressurized, the joint becomes subject to rotational forces (Fig. 3.9b) which open up the secondary seal. The joint no longer has seal redundancy. The figure also shows that there is a high probability of secondary seal failure after the first 330

Figure 3.9   Solid booster rocket showing loss of backup seal in the lift-off phase.

milliseconds of rocket firing. Seven astronauts died from just such a design inadequacy in the *Challenger* accident in 1986.

Column 5: Failure mechanism (cause). The information on causes has two purposes. First, identifying the root cause helps engineers eliminate it by redesigning the product. Second, identifying the stress (thermal, shear, tensile, or compressive, for example) that causes the failure makes it possible to validate a test program that will assure robustness of the component for this stress. In some cases, identifying the stress may not be possible.

Column 6: Effects of failure. The effects of the component failure on the product should be considered all the way to the user level. To do that, the effect has to be traced to higher-level assemblies. For example, a failure of the connecting rod in an automobile engine will have some effect on the piston performance which in turn will result in automobile failure. The purpose of this column in the FMECA is to get some idea about the risk.

Column 7: Interface effects. This information is not required in the normal FMECA but is important added information to the FMECA analyst. It shows how the end product interfaces with software or with other products. If the effect of the failure under consideration causes the software or other product to malfunction, then the risk must be evaluated. A possible redesign may be considered.

Column 8: Criticality. The automotive industry and many commercial organizations follow the *risk priority number system*. In this method, each failure is rated in three categories and the ratings are multiplied together to arrive at an index called the *risk priority number* (RPN). The three categories are severity, frequency, and detectability. The rating scale in each is from 1 to 10, where 10 is the worst rating. The detectability category has a special meaning. If an incipient failure is detectable before it becomes a failure, then the detectability rating is low. The worst rating of 10 implies absolutely no advance warning of a catastrophic failure. For example, an automobile that goes into sudden acceleration without any warning would be rated 10. This feature makes designers aware of the need for a safety or warning device.

In this system, RPN can be anywhere between 1 and 1000. Note that one has to be cautious about a potential catastrophic failure being masked by low ratings on frequency and detectability. For example, a failure with a rating of $10 \times 2 \times 2$ has a rating of only 40 and may appear to be unimportant. One must make sure that every failure with a severity rating of 9 or 10 is assessed for safety no matter what the overall rating.

Another popular system of criticality ranking is by the magnitude of failure rates only. This does not necessarily reflect safety risks.

The U.S. government uses a system slightly different from the RPN system. The criticality is related to risk analysis. The two primary components of the risk are severity and frequency of the failure. Whether a team takes an action or not depends on this information. There are several approaches to how the information is used. The Department of Defense assigns qualitative ratings to severity and frequency. The severity ratings are

I—Catastrophic failure or loss of system

II—Critical failure or major loss to system

III—Minor failure or minor injury

IV—Incidental failure

Similarly, the frequency ratings are from A through E:

A—Very likely to fail

E—Almost impossible to fail

The inherent weakness in this system is that if there are 800 category I failure modes, it is difficult to rank them according to priority.

Column 9: Corrective action. This column is the one which is often misused. The purpose is to eliminate category I and II failures by design. Many analysts instead recommend inspection and testing (which may be acceptable for category III and IV failure modes). Many times a fault-tree analysis will help in coming up with the appropriate design change. If the failure is not eliminated by design, then this column should indicate the rationale for handling this failure any other way. To assure that at least a sincere attempt is made to improve design, some companies provide reevaluation of severity, frequency, and detectability as in Fig. 3.10.

Column 10: Item close sign-off. Often the analysis is done but no action is taken. This column is introduced as an optional column to assure the important failure modes are taken care of. In some countries, criminal actions can be brought against engineers and managers for ignoring safety issues.

FMECAs have been modified from time to time. They can be used

Part name/ part number	Potential failure modes	Causes (failure mechanism)	Effects	Risk priority rating				Recommended improvement	Improved rating			
				Sev*	Freq	Det	RPN		Sev	Freq	Det	RPN
Pipe	Leakage in pipe	1. Corrosion	Loss of Freon	4	3	8	96	Use stainless steel pipe.	4	1	8	32
		2. Temperature cycling	Loss of Freon	4	3	8	96	Monitor temperature with thermocouples.	4	2	2	16
	Leakage at the joint	1. Cumulative fatigue	Loss of Freon	4	4	8	128	Monitor vibration with acceleration	4	2	2	16
		2. Poor soldering	Loss of Freon	4	4	5	80	Update process FMEA	TBD	TBD	TBD	TBD
Valve	Sticky, intermittent	1. Dirt or foreign objects	Loss of control of temperature	7	3	8	168	Electronic redundant valve action and fault identification	7	1	1	7
	Stuck open	1. Component wearout	Loss of control of temperature	7	2	8	112	Electronic redundant valve action and fault identification	7	1	1	7
	Stuck closed	1. Component failed	Loss of control of temperature	9	2	8	144	Electronic redundant valve action and fault identification	7	1	1	7
		2. Component expansion and contraction	Loss of control of temperature	9	2	8	144	Electronic redundant valve action and fault identification	7	1	1	7

* Severity ratings 8 to 10 request special effort in design improvement regardless of RPN rating.

Figure 3.10 FMECA with reevaluation of risks.

for different purposes as indicated earlier. Actually the FMECA should be updated as the product develops.

Cautions on FMECA. One of the features that creates confusion about criticality is fault tolerance. A system can tolerate a fault if there is a redundant backup. Ordinarily a reliability engineer will regard the failure of a component in redundancy as not critical. A safety engineer, on the other hand, may call it critical because, once the primary component has failed, the system no longer has a backup. The system at NASA overcomes this problem by adding R for redundancy in the criticality rating—for example, IR, IIR—where a redundant element is present. This is an indication to the designer to consider providing a fault alarm for the failure of the primary component. The alarm will call for restoring the redundancy.

A bigger problem is that a vast majority of companies violate the first principle of FMECA, that an analysis be done by a designer and a nondesigner together. This gives a more thorough critique of the design. Usually, a reliability engineer does the analysis alone. Except in rare cases, this is a waste of valuable resources because a reliability engineer alone is not qualified to judge the details of design. As a team, the design engineer and the reliability engineer can truly develop a highly reliable product.

New applications of FMECA. The future role of FMECA will be bigger than it is today. Signs of increasing acceptance abound. The mass transit system in Atlanta uses FMECA in writing maintenance procedures. At a Polaroid plant, FMECA is used for training repair personnel. At a candy mass production plant, maintenance technicians use the tool for solving engineering problems. Its role will expand to maintainability analysis, process control, testability analysis, and software reliability analysis. Some commercial companies are already performing process FMECA. However, most aerospace and defense contractors are still unaware of the benefits of applying FMECA to manufacturing processes.

FMECA is a versatile tool. For software, it can identify missing requirements in the specifications and point out sneak paths in the code. For maintainability analysis, FMECA can identify critical microprocessor paths that must be tested (but which are often untestable).

The author has performed FMECA at various levels and in different forms. The hierarchy in Fig. 3.11 has been found highly productive for high-technology products.

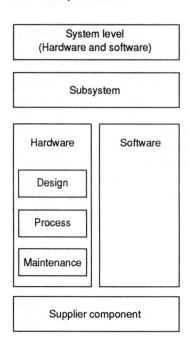

Figure 3.11   FMECA hierarchy.

3.2.7 Worst-case analysis

Worst-case analysis can be done in many ways, although the objective always is to determine whether a product will still work under the worst possible conditions. For electronic circuits, effects on the entire circuit are evaluated for component values at the lowest and highest ends of the tolerance. This also allows prediction of drifts in the engineering parameters. For mechanical components, similar evaluations are made to determine the effects on interchangeability and on the overall assembly performance. In all cases, the worst effect of all the possible environments can also be evaluated.

Worst-case analysis can also be used for quickly determining the effects of engineering changes. For example, an engineer who is not sure whether a new tolerance on a component is adequate can test some components at the low end of the tolerance and some at the high end. If both ends give satisfactory performance, there is perhaps no need to test components in the middle of the distribution. This form of testing will give much higher confidence with fewer components tested. Figure 3.12 shows such an approach.

Worst-case analysis should go beyond tolerance analysis, however. Failures can be caused by electrical noise, race conditions, storage in a warehouse, electrostatic discharge, electromagnetic interference, and

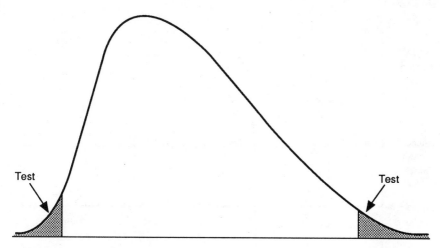

Figure 3.12   Worst-case testing.

human errors. The analysis should be done for static as well as dynamic conditions.

3.2.8 Other analysis techniques

Fault-tree analysis and sneak circuit analysis are covered in Chap. 5. The robust design approach and the quality function deployment technique are covered in Chap. 6. Statistical analysis of failure distributions is covered in Chap. 2.

3.2.9 Design improvement approaches

Derating and fault-tolerant design are the most frequently and gainfully used design improvement techniques. Zero-failure design is covered in Chap. 5 and modular design is covered in Chap. 4.

Derating.   This technique became popular in electronic design, but it is used to provide design safety margins in other fields as well. The concept can be explained from an S-N diagram, which is a graph of stress (y axis) versus number of cycles to failure (x axis). For a given stress, the diagram gives the cycles to failure. Today, the x axis may be plotted in miles or hours to failure, as in Fig. 3.13a, or in the traditional units of number of cycles. On log-log paper the diagram is usually a straight line plot with a knee at some low value of stress, as in the figure. This diagram shows that if the stress on the component is lower, the component will last longer. If the stress is below the knee level, the component can last almost forever (100 million cycles is considered an infinite life for practical purposes). The reduction of

(a)

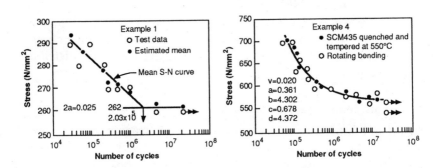

The proposed S-N curve may be expressed as

$$y = a\{ \sqrt{(x - .d)^2 + c^2} - x \} + b$$

where $y = \log S$, $x = \log N$ and a, b, c and d are constants to be optimized. When $c = 0$ this equation yields the simplest bi-linear relation composed of inclined and horizontal straight lines; the slope is then given by $-2a$, the knee position by d, and the fatigue limit by $b - ad$.

(b)

Figure 3.13 (a) S-N diagram for a high-voltage component based on tests conducted by the author. (b) S-N diagrams reported by Nishijima. (Source: T. Tanaka, S. Nishijima, and M. Ichikawa, Statistical Research on Fatigue and Fracture, Elsevier Applied Science, New York, 1987.)

stress is expressed as a *derating factor* and can be roughly equated with reliability level:

Derating factor	Reliability level
50% of maximum component rating	High reliability
30% of maximum component rating	Very high reliability

The 30 percent rating usually is equivalent to a stress level below the knee. At this stress the component in Fig. 3.13a, a high-voltage cable, shows a life of more than 30 years.

Caution: There are some exceptions to the above rule. For example, an aluminum electrolytic capacitor will have lower reliability if the stress falls below a certain threshold. Frozen batteries may not work at all.

Fault tolerance.   Fault tolerance usually implies two devices (one a backup to assure functional performance). However, for measurement or computation in such applications as aircraft instruments or software control, an odd number of devices may be used, and a decision is made by majority vote. This type of odd redundancy is often called *voting redundancy*. Whether two or three or even more are required will depend on the savings assessed from the life-cycle cost projection for each option.

Redundancy may be *active* or *standby* in nature. In active redundancy, the primary and the backup devices both operate at the same time. They may share the load. In standby mode the secondary device is switched on only when the primary device fails. The advantages and disadvantages are compared below.

I. Active redundancy
   A. Advantages
      1. The system is automatically derated, since each device is sharing the load. This increases the life of the devices.
      2. Availability is increased because the failed device stops sharing the load and gets immediate attention by repair personnel.
      3. The reliability of an individual device may be inherently higher when it is in the operating condition at low stress than when it is in storage.
   B. Disadvantages
      1. The cost of two devices is higher than that of one. This disadvantage is not significant if the whole equipment does not need to be duplicated. Many times a subassembly or

just a few components are required to be redundant. For example, a company provided a redundant component on a circuit board instead of the whole board. By the same token, one could include a redundant circuit board instead of duplicating the whole black box. (One medical company provided a redundant fan to keep circuitry cooler instead of providing a redundant black box.)
  2. More maintenance may be required because of additional devices.
II. Standby redundancy
  A. Advantages
    1. The backup can take the full load when the primary has failed.
    2. Backup maintenance can be performed without disrupting the system.
  B. Disadvantages
    1. Backup equipment may fail to turn on. Switching reliability is important.
    2. Backup may fail without maintenance personnel knowing it. Therefore, standby systems require frequent monitoring or automated diagnostic checks.
    3. Storage and the nonoperational standby condition may introduce additional failure modes.

As noted, using only one backup device may not be satisfactory where numerical data are involved. For example, if one computer gives a wrong output because of a malfunction, and its backup gives the right answer, the user may not know which one is right. If there were three computers, the two right ones could vote by giving a matching output. This is called *two-of-three voting redundancy*. On the NASA space shuttle that exploded in 1986, there were five computers to allow for sufficient voting in case one or two of them fail totally. This approach is often used for checking errors caused by noise or line-voltage transients. In general, if k out of n devices must work, the model is called k-of-n redundancy. For computation of reliabilities, refer to Chap. 2.

Caution: Redundancy is not always the answer if the redundant devices have a serial element that limits reliability, such as a common power supply. The reliability of the power supply then becomes critical. The system may require an independent power supply for each device. Analysis of such problems is called *common-cause analysis*. Usually, fault-tree analysis and minimum cut set analysis are required. This information can be easily obtained by a glance at the block diagrams for system reliability. This approach is covered in Chap. 5.

3.3   Reliability in the Manufacturing Phase

The high early failure rate of the bathtub curve is mostly from manufacturing discrepancies, although some manufacturing discrepancies can increase failure rates in the useful life and wearout portions of the curve as well. Figure 3.14 shows the effect of manufacturing screening on field failure rates.

The objective of designing for reliability in manufacturing is to control the manufacturing-related causes of failure. To identify these causes, an FMECA for the manufacturing process should be done before the process is installed. This is covered in Chap. 6, "Quality Assurance." Process FMECA shows the causes of process discrepancies during the design of the process. Many additional causes of failures will be discovered during a pilot run. It is at this time that process control is needed most, before any defective product goes out the door. As an example, consider an operator who takes too long to solder a joint, thereby producing a poor connection. One solution to the problem would be to study how long (and short) a time the solder gun should be kept on to produce a reliable joint. This would establish the

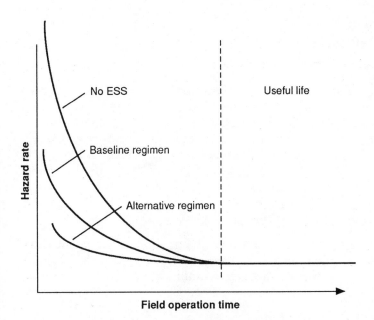

Figure 3.14   Effect of screening on field failures. (*Source: USAF R&M Process, Handbook SAF/AQ, U.S. Air Force, Washington, D.C., 1987.*)

tolerances for the solder time. If adhering to soldering tolerances still does not produce reliable joints all the time, then it may be necessary to change the soldering process itself, or to use a different process altogether.

The action required to ensure reliability is not necessarily the same action required for traditional quality control. Not all discrepancies or defects lead to low reliability. For example, these defects may degrade quality but not reliability:

The wrong shade of a color

A light dent on the surface of a casting

A scratch on the paint.

A poor surface finish

The wrong plating on screws

However, these defects or flaws usually reduce reliability:

A poor weld

A cold-soldered joint

Leaving out a lockwasher

Using the wrong flux

Not cleaning the surfaces to be joined

A large dent on a spring

An improper crimp on a wire joint

Every failure in a test and every nonconformance to specifications should be evaluated for effects on reliability. This area has received little attention, but it has the highest influence on product costs. If manufacturing defects are controlled, then burn-in and screening test costs can be drastically reduced. One-hundred percent screening of parts may not be required. Only an audit may be required to make sure no new errors are introduced.

3.4   Reliability in the Test Phase

The test phase of reliability starts in the product development stage and continues in the production phase at less rigorous levels. The following reliability tests are done as a minimum:

1. During design
   a. Reliability growth tests
   b. Durability tests
   c. Reliability qualification tests (same as qualification tests during manufacturing)

2. During manufacturing
   a. Burn-in and screening
   b. Failure-rate tests, also called MTBF tests

3.4.1 Reliability growth testing

In complex equipment, good designers make few mistakes while hasty ones leave many problems unsolved. To catch such problems, the *reliability growth* concept was developed. The concept calls for identifying problems and solving them as the design progresses. To accomplish this, a closed-loop corrective action procedure called TAAF, for *test, analyze, and fix*, is used. The idea is that, if environmental tests are done in early design, the product should mature faster and achieve higher reliability, in what James T. Duane called a *learning curve approach*. In 1964, Duane published a simple model that showed cumulative failure rate linearly related to cumulative operating hours on log-log paper.[9] Other models have appeared since then (see Mil-Hdbk-189),[10] but because of its simplicity the Duane model has prevailed.

The *Duane model* is characterized by the equation

$$\text{Cumulative } \lambda_c = kt^{-b}$$

This can also be written as

$$\text{Cumulative MTBF} = \frac{1}{k}t^b$$

where b = slope
  k = starting failure rate
  t = cumulative test time

When logarithms of both sides of the equation are taken, the equation becomes linear:

$$\log \lambda_c = \log k - b \log t$$

Data on three groups of vehicles showed the following relations:
Group A:

$$\lambda_c = 5.47t^{-0.6632}$$

Group B:

$$\lambda_c = 0.719t^{-0.3595}$$

Group C:

$$\lambda_c = 0.562t^{-0.3189}$$

Today most companies track MTBF rather than λ since it is easier to communicate whole numbers instead of fractions (assuming constant failure rate). The Duane model shows that the design fixes are effective and implemented before the next test. If many engineering changes have been made, credit for them should be taken by computing instantaneous MTBF, which can be derived from the cumulative failure rate equation as follows. λ_c can be expressed as the number of failures $N(t)$ within time t ($t > 0$) divided by t, or

$$\lambda_c = \frac{N(t)}{t} \quad \text{and} \quad N(t) = \lambda_c t = kt^{(1 - b)}$$

The change per unit time of $N(t)$ is the instantaneous failure rate λ_i which when $t \to 0$ is

$$\lambda_i = \frac{d\,[N(t)]}{dt} = k(1 - b)t^{-b} = (1 - b)\lambda_c$$

Since instantaneous MTBF is the reciprocal of the instantaneous failure rate, the instantaneous MTBF is

$$M_i = \frac{1}{\lambda_i} = \frac{1}{(1 - b)\lambda_c} = \frac{\lambda_c}{(1 - b)}$$

where M_c is the cumulative MTBF. Figure 3.15 shows cumulative and instantaneous MTBF curves.

AMSAA model. The Duane model assumes constant reliability growth. The growth actually can accelerate or decelerate, depending

(a)   (b)

Figure 3.15   Reliability growth models. (a) Duane model; (b) AMSAA model.

on the competency of the design team and the number of engineering changes made after each test. Concern over variable reliability growth led the U.S. Army to a new path. The Army Material Systems Analysis Activity (AMSAA) model assumes that the reliability growth is a step function from one test phase to another, as shown in Fig. 3.15b. At a workshop of the Institute of Environmental Sciences in March 1988, the AMSAA model was recommended by the Army and the Air Force Rome Air Development Center. The AMSAA model is broader in application, and the Air Force step model[3] is not greatly different. This approach reflects better fit. For details of the AMSAA model the reader is referred to Mil-Hdbk-189.[10]

Regardless of the reliability growth model, two fundamental, related questions must be answered in a growth program: What should be the starting MTBF? What should be the growth rate?

Traditionally, projects have used a starting MTBF value between 10 and 30 percent of the goal. Proponents say that this is realistic, according to experience. An Air Force manager, however, favors going to a higher starting MTBF, since starting at a low percentage indicates a need for much more design work. He prefers to enter the growth test with a well-engineered design. This author agrees with him. Reliability is a mature science now. Commercial manufacturers would be out of business if their reliability grew so slowly. Most engineers in aggressive commercial sectors try to achieve more than the 100 percent MTBF goal at the outset. Some succeed, some do not. One manufacturer of Winchester disk drives had an MTBF of 4000 hours when a competitor came up with an MTBF of 5000 hours in a drive at half the price. To compete head-on, the manufacturer went into a reliability growth program with a starting MTBF of 6000 hours and a goal of 10,000 hours over 3 years. But the competitor introduced a product with 20,000-hour MTBF in 1 year—at a lower price, too!

The answer to how fast the reliability should grow depends on many factors. Some of the factors are discussed in the following sections.

TAAF versus ATAF. The truth is that reliability growth in a complex product can be faster if most of the design changes are made because of proactive analysis instead of reactive testing. The intent of reliability growth programs is to encourage a thorough analysis before testing, but usually the sequence is reversed. Microanalysis takes place after testing instead of macroanalysis before. FMECA and fault-tree analysis are often used too late, if used properly at all. Good design principles call for detailed analysis first so that most of the design improvements are made while the design is still on the drawing board. It is much cheaper to make changes at this stage than after the hardware and tooling have been fabricated. What is needed is ATAF (analyze first, then test and fix) instead of TAAF. In other words, always perform FMECA before implementing reliability growth tests.

Reliability growth through accelerated testing. The commercial sector has fostered reliability growth for many years without calling it by that name. When the author worked in the commercial sector, he had to accept (out of necessity, of course) the challenge of improving the MTBF of a product line tenfold in one year (1000 percent). The only way to succeed was to resort to highly accelerated but valid tests, described later in this chapter.

Without support from design engineering, this product line would have vanished from that company. The company did achieve a major breakthrough that enabled it to regain all its lost market share. It even guaranteed the product for 15 years at no charge to the customer. As a result of the improvements, the return on investment for that business increased from 17 to 71 percent!

Companies like Cray Research, Apple Computer, IBM, Polaroid, Hewlett-Packard, Motorola, and Intel have stunned the industry with reliability growth many times. Many agree to the need for higher acceleration in testing, but not enough is published about it.

Controlling warfare over failure causes. Most projects are plagued by disagreements over the causes of failure. Software engineers tend to ignore hardware failures. Bob Cruickshank of IBM Federal Systems makes this point clearly, with a chuckle: "How many programmers are needed to change a light bulb? None! Because this is a hardware problem!" Hardware engineers often do the same; they work independently of the software engineers. The systems then inherit interface problems that sneak up at a much later stage. Reliability predictions are then shaky, to say the least. As long as major engineering errors and single-point failures remain in the system, reliability growth is hard to control. In addition, the system cannot be designed to fail gracefully.

Reliability growth testing is necessary, otherwise design and management may overlook reliability problems. The growth management clause in contracts gets attention up front during R&D. Then more models will come, such as software growth models, maintainability growth models, and availability growth models. As reliability lecturer Allan S. Golant says of the Duane growth model, "Its inherent simplicity makes it the preferred method if communication with your management is the prime requisite."

3.4.2 Tests for durability

Durable life can be defined as the life to the point at which the wearout starts to accelerate, where the third region of the bathtub curve starts. To establish this point, the data from fatigue testing only are considered. (The presence of infant mortality failures will corrupt the data.) Depending on the severity of failure, this point is selected so that the cumulative failure probability is less than 1, 5, or 10 percent. The lives are then commonly referred to as B_{01} life, B_{05} life, or B_{10} life.

Many think that MTBF is a measure of durable life, but there is absolutely no relation between the two. As noted earlier, MTBF is strictly a function of failure rate, while durable life is a function of length of the bathtub. Two bathtubs can have the same failure rate (therefore the same MTBF) but different lengths.

Some consider durable life as the mean life. This concept has little value in the reliability literature, because by the time the mean life is achieved, roughly 37 to 68 percent of the products would be expected to fail, depending on the probability distribution. Who wants a product with a 63 percent chance of failure? It is amazing how much business is conducted on the basis of advertisements of mean life without manufacturers or customers realizing what it means. A pacemaker manufacturer told doctors that the battery in the product had an average life of 4 years. The doctors thought each battery would last 4 years, and so their patients were told. No wonder that many patients sued the doctors and the doctors sued the manufacturer. As might be expected, a high percentage of the patients found that their batteries failed before 4 years, some in less than 6 months. Since the batteries did not give advance warning of failure, some patients died. Such cases abound. Think of a light bulb. When you buy a lamp with an average life of 2000 hours, do you expect each lamp to last 2000 hours? How many people know that only one out of every two such lamps may last more than 2000 hours? The same is true for products ranging from cadmium batteries to satellites; it should not have been a surprise that a weather satellite with a mean life of 5 years failed in 2 years.

Durability is assessed in many ways. Usually some empirical model is used. This chapter will cover several models which are often used. But first, discussion of accelerated life tests is in order. Accelerated tests are necessary because testing life under normal conditions would take months or years.

Accelerated life testing. The purpose of accelerated testing is to duplicate the field failures by providing a harsher—but nonetheless representative—environment. The product is expected to fail in the lab just as it would have failed in the field—but in much less time. Then an empirical relation can be used to predict field life from lab life.

Before any prediction of life, it is necessary to make sure certain requirements are met. The test should not create failure modes that the customers will never see. The four basic requirements, based on the author's long experience, are

1. The predominant failure mode under field stress and under accelerated stress should be same. This is an indication of the validity of the accelerated test profile. The stress (such as voltage or temperature) that creates such a failure is called *dominating stress*.

2. The engineering properties of a material under accelerated stress should be the same before and after the test, except at the point of failure. Properties such as lattice structure or grain structure, for example, can usually be defined by a metallurgist. This means that the range of stress application is within the elastic limits of the material.

3. The shape of the failure probability density function at rated and higher stress levels should be the same. This is demonstrated by the curves having the same slope on a Weibull graph. See Chap. 2.

4. The Weibull characteristic life should be repeatable among production lots, within 5 percent.

As long as the above four conditions are satisfied, one can select higher stresses to reduce test time. The most popular approach is to construct an *S-N* diagram as a test model. Figure 3.13*a* shows an *S-N* diagram for a high-voltage connector made from the same material as high-voltage cables. The customer uses the connector at 15 kV, but it cannot be practically tested at that voltage (it would take 15 to 30 years to fail). The voltage therefore was increased to 40 kV, and, since one unit cannot be representative of production, a random sample was tested at that voltage. Usually Weibull characteristic life at 63.2 percent failure gives a good, representative point for the *S-N* diagram. Similarly, Weibull characteristic lives were plotted at 50 and 60 kV. These three Weibull characteristic lives fell on a straight line on log-log paper. The characteristic life at 70 kV did not fit on this line, indicating that that stress was too high for the material. At this level, the failure mode also changed, so it was clear that one of four principles cited has been violated. Once the *S-N* diagram is constructed, the graph can be projected to give a life estimate at 15 kV.

Accelerated test models for durability prediction. The two most popular models in use are the inverse power law model and the Arrhenius model. The inverse power law model is a versatile model for many different mechanical and electrical stresses. It states that life is inversely proportional to the stress raised to the power N, where N is the acceleration factor derived from the slope of the S-N diagram. It is computed from the equation

$$\text{Slope} = \frac{-1}{N}$$

(The slope in the graph is negative.)
The model can be written as

$$\frac{\text{Life at rated stress}}{\text{Life at accelerated stress}} = \left[\frac{\text{accelerated stress}}{\text{normal stress}}\right]^N$$

For the product in Fig. 3.13a, the value of N was found to be 8.2 from the slope of the S-N diagram.

How long should the test last? This too can be found from the inverse power law. For the product in Fig. 3.13a,

$$\frac{30 \text{ years} \times 365 \text{ days} \times 24 \text{ hours}}{\text{Test hours at accelerated stress}} = \left[\frac{60}{15}\right]^{8.2}$$

The test time at the accelerated stress is therefore 4.01 hours.

Caution: Make sure all four accelerated test principles are valid at each accelerated stress.

Arrhenius model.   This model has been widely misused. The intent of the model is to predict life at normal temperatures from accelerated test data. However, it applies only to failures from chemical reactions, or metal diffusions, which are accelerated by temperature. It does not apply to failures generated by burn-in and screening for infant mortality.

The Arrhenius model can be applied to insulating fluids, lubricants, battery cells, plastics, and some failure mechanisms of electronic components. But care should be exercised when applying this model to electronics because most so-called electronic failures are really mechanical failures. The Arrhenius model is given by

$$\text{Life} = Ae^{E/kT}$$

where A = empirical constant
T = temperature, kelvins (for electronics, the junction temperature; for fluids and other products, the hot spot temperature)
k = Boltzmann's constant = 8.62×10^{-5} eV/K
E = activation energy

Arrhenius graph paper (Fig. 3.16) has a special grid for plotting the Arrhenius equation as life (time to failure) versus temperature in degrees Celsius.

The value of E is obtained from the relation

$$E = -k \times \text{slope}$$

The value of A is obtained by selecting any point in the graph and using its x and y coordinates (life, temperature) and the slope to compute A from Eq. (3.3).

Once these values are known, the model can be used for testing at high temperature to predict life at lower temperatures. Extrapolation can be done toward lower temperatures, but not below the knee temperatures and certainly not below freezing temperatures. Comparison of life at two temperatures from the graph will lead to the following equation, which is used by some organizations:

$$\frac{L_1}{L_2} = e^{[(E/k)(1/T_1 - 1/T_2)]}$$

Many in the electronics industry assume that the life and failure rates are inversely proportional. (The author disagrees. See the Sec. 3-1 argument on life versus MTBF.) If the assumption is valid,

$$\frac{\lambda_2}{\lambda_1} = e^{[(E/k)(1/T_2 - 1/T_1)]}$$

Caution:

1. The Arrehenius model is an empirical model. It should be applied to failure rates with proper understanding. Chemical reaction rate and mechanical failure rates on electronic devices are not related. The failures may result from nonchemical causes such as a transient voltage. Certainly, infant mortality failures do not belong in this model.

2. Some practitioners determine the value of E only once and apply it across the board to other components. This practice is not recommended. Failure rates vary from one component to another, even though they may have similar designs. Units with the same part number manufactured in two geographical locations can have two different activation energies.

Models for specific applications.   A component may go through cyclical or varying loads, and the models just described may not apply. However, the literature contains several models for such components. Among them are Goodman diagrams[11] and Miner's rule.[12,13] Chapter

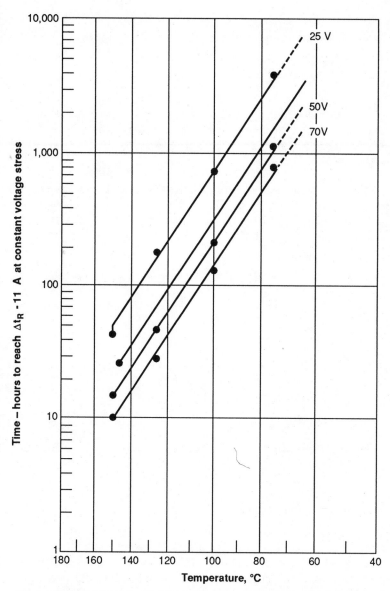

Figure 3.16   An Arrhenius plot for a discrete semiconductor device. (*Source: Erwin A. Herr, "Reliability Evaluation and Prediction for Discrete Semiconductors," IEEE Transactions on Reliability, August 1980.*)

5 covers discussions on safety margins. An illustration of Miner's Rule follows below.

Miner was a design engineer at Westinghouse. He postulated that a component subjected to different cyclical loads accumulates damage and that failure can be expected when the summation of incremental damage from each load equals unity. Failure is likely when

$$\frac{C_1}{N_1} + \frac{C_2}{N_2} + \frac{C_3}{N_3} + \cdots + \frac{C_i}{N_i} = 1$$

where C_i = number of cycles applied at stress S and N_i = capability of the material in cycles at S, determined from an S-N diagram.

This equation shows that tests to establish an S-N diagram are required for this approach.

Example 3.3   A component is expected to see the following stresses during its lifetime:

150,000 cycles at 2600 lb/in^2
100,000 cycles at 2900 lb/in^2
25,000 cycles at 3200 lb/in^2

The capability as determined from the S-N diagram at these stress levels is

400,000 cycles at 2600 lb/in^2
250,000 cycles at 2900 lb/in^2
190,000 cycles at 3200 lb/in^2

From Miner's rule, the cumulative damage is

$$\frac{150,000}{400,000} + \frac{100,000}{250,000} + \frac{25,000}{190,000} = 0.907$$

Since the cumulative damage is less than 1, failure is not expected.

Caution: Studies have shown that failure takes place in the range of 0.7 to 2.3 cumulative damage value.

3.4.3   Testing for low failure rates

For simplicity, MTBF has been used as a measure of failure rate even though it is based on the assumption that failure rate is constant. In practice, failure rate is rarely constant. Infant mortality failures may be present in the field for any product, and for mechanical products the failure rate is almost always an increasing function of time. But industry marches ahead with the assumption of constant failure rate. Industry should accept failure rate for what it is: it can be increasing, constant, or decreasing with time. In such situations, Weibull analysis may be the appropriate choice for failure tests. MTBF should be used only when the failure rate is truly constant. Failure rate tests are done both for reliability qualification of new products and for reliability demonstration during production.

If the failure rate is increasing or decreasing with time, the times at which the failure rate should be reported must be specified. The Weibull failure rate model from Chap. 2 can be used. The goal should be to reduce failure rate to a small constant value during the useful life.

Reliability qualification tests. Reliability qualification is done on production devices but should also be done on prototype models. It should be an accelerated test if the failure rate for the entire expected life must be verified. Many companies are not quite sure how to accelerate the tests (see Sec. 3.4.2 for guidance). Others are afraid to overstress beyond the published ratings. Most companies qualify only for a portion of the bathtub curve.

Qualification tests consist of stressing the product for all the expected failure mechanisms. The test can be stopped if there are no failures during the expected life. The test conditions for qualification are generally severe. Some may not be performed for production demonstration tests, such as thermal shock, impact, and constant acceleration. Some special stresses may have to be applied to duplicate field conditions. For example, a bias voltage may be needed to create the worst electron migration in integrated circuits. Similarly, tensile and compressive forces may be needed to stress a spring for a mechanical apparatus. An impact load may be required for an aircraft structure.

Example 3.4 A high-voltage fuse was qualified for 30-year life under these accelerated test conditions:

1. Environmental reliability
   a. Lightning surges and continuous dielectric stress
   b. Environmental such as moisture, oil, dust, and fungi
   c. Ultraviolet rays from sunlight
   d. Current in the customer's system which could be continuous or transient during load application or fault interruption
2. Mechanical reliability
   a. Vibrations
   b. Mechanical switching
   c. Magnetic forces
3. Thermal reliability
   a. Operating temperature
   b. Temperature cycling
   c. Rate of change of temperature
4. Operational reliability
   a. Physical operation of components
   b. Installation practice
5. Interface reliability
   a. Interface with customer's system
   b. Interfaces between components, such as soldered, welded, and mechanically connected joints and connectors for electronics
   c. Interface between subassemblies
   d. Interface between subassemblies and final assembly
   e. Interface with software

6. Use reliability
   a. Field instructions
   b. Repair practices
   c. Maintenance practices
   d. Logistics support practices
   e. Handling and storage
   f. Tools

Reliability demonstration tests, on the other hand, usually require stressing during only a portion of the useful life. They do not duplicate all the tests done during product qualification. Since reliability demonstration tests are not very severe, qualification tests should be done at preset intervals. For electrical relays, for example, Mil-Std-454 calls for qualifying every 2 years.

Caution: The interval for qualification tests should be carefully chosen. For example, the author at one point had to investigate a relay failure that took at least 11 lives. The relay was qualified for leakage of salt water every 2 years. The leak tests in production were not very severe, and defective relays went out to the field. If the qualification tests had been done more frequently on a few units, most of the defective units would have been detected earlier.

Qualification test data should also be analyzed to determine whether the product meets the qualification criteria. The basic criteria are

1. No infant mortality failures should go to the customer.

2. Failure rate should be below a target percentage over the useful life (such as less than one failure in 10 years or, for an automobile, 100,000 miles).

Example 3.5   The data in the table below were recorded for an accelerated life test of Winchester disk drives for mainframe computers. The sample of 16 drives was drawn from a finished-goods warehouse. Each hour of accelerated testing represents approximately 0.4 month in the field. The quality criteria were that customers should not experience early failures and that the failure rate during the expected 7-year life should be less than 0.5 failures per 100 disk drives. Assuming the test environment is a very close duplicate of field environment, is this product qualified?

Failure rank number	Failure mode	Hours to failure
1	Servo	7
2	Read/write	12
3	Power supply	49
4	Read/write	140
5	Head crash	235
6	Head crash	260

Failure rank number	Failure mode	Hours to failure
7	Servo	320
8	Head crash	320
9	Head crash	380
10	Head crash	388
11	Personality	437
12	Microprobe	472
13	Head crash	493
14	Read/write	524
15	Head crash	529
16	Head crash	592

Since the Weibull distribution fits over 90 percent of life test data, it will be appropriate to plot the tabulated data on a Weibull graph as in Fig. 3.17. The graph shows that part of the data fits on one slope and the remaining data on another slope. This is typical of a product infested with manufacturing-induced "weak" failures. The first slope reflects infant mortality failures, and this should be verified with failure analysis tools. The weak failures are often a result of excess variability in quality; the failure mechanisms are usually obvious: cold solder joints, loose lockwashers, voids, and nonhomogenous material. The second slope shows that the product goes straight into wearout phenomena. There are hardly any random failures in the laboratory for mechanical products. In electronic products, of course, three slopes are possible, one for each region of the bathtub curve.

The plot shows that this product is not qualified for market. Since these units were drawn from the finished-goods warehouse, the infant mortality failures should not have been present. Apparently, the screening was insufficient. The graph shows the change in slope at 200 hours. The failures during the initial 200 hours are likely to be from production problems, and this should be verified. The knee therefore signifies that the product can be shipped after every unit goes through an additional screening of at least 200 hours. The alternative is to solve manufacturing problems to eliminate causes of infant mortality.

The second criterion, that failure rate be below a certain target value, cannot be determined from the data. In the data, the "lemons" (infant mortality failures) and the "pineapples" (inherent design failures) are mixed. When two such populations are mixed, prediction is risky. In fact, since the infant mortality failures change their personality from day to day, prediction is almost impossible.

Assuming the causes of infant mortality failures are eliminated (or an expensive screening program is in place), the data would have been without the first four failures. The plot of data without infant mortal-

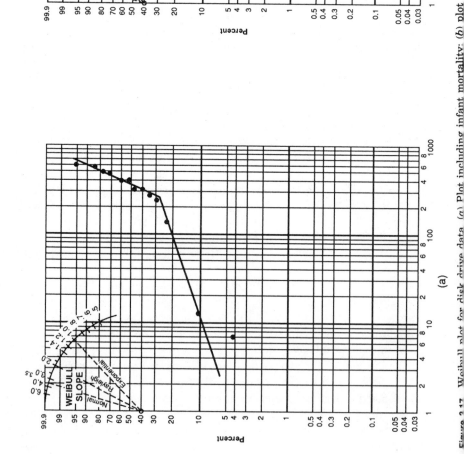

Figure 2.17 Weibull plot for disk drive data. (a) Plot including infant mortality; (b) plot

ity failures (sample size 12 instead of 16 as in Fig. 3.17b) shows that the expected failures in 7 years (84 months = 210 accelerated hours) is 5 percent, or 5 out of every 100 disk drives. This is not acceptable, since the goal is to have only 0.5 failure per 100. The product therefore fails to qualify.

Caution: Many users apply Weibull analysis at the system level, where there are mixed failure modes. Weibull analysis is intended for fatigue experiments, which can be applied only at component level, or for an assembly where a same failure mode appears repeatedly, such as failures in solder joints, welded joints, and a simple assembly of two mating parts.

It is easy to get a straight line on a Weibull graph at system level. But the slope of the line will change from production lot to production lot (unless production is so highly mechanized that the same defects occur all the time). The question then is: How does one predict field reliability at system level? The answer is: One doesn't. At system level one can only postulate, because many failure mechanisms are neither part of reliability model nor the same every time. Some examples of these variables are bad wire connections, miswiring, electric power transients, electrostatic discharge, electromagnetic interference, poor maintenance, failure to follow instructions, human errors, and software errors.

Weibull software programs should also be used with great caution. Most of them fit a line through two or more distributions of the bathtub. Statistically, a model can be fitted to mixed populations (some fit a log-normal model) but may violate basic engineering principles for solving problems. The critical failure mechanisms may no longer be visible. A single model for mixed populations may be a good indicator of the past but is an uncertain indicator of the future.

Reliability demonstration tests. Under less severe conditions, the production lots are accepted on the basis of failure rate tests. Most tests are operated under the assumption that failure rate is constant. Therefore, MTBF is the common measurement. Department of Defense handbook H108[15] contains three types of sampling plans:

- *Failure truncated:* Test is terminated at a predetermined number of failures.

- *Time censored:* Test is terminated at a predetermined time.

- *Sequential testing:* Covered in Mil-Std-781.[16]

In these plans the customer and the supplier decide on four factors: the producer's risk (chance of rejecting a good product), the consumer's risk (chance of accepting a rejectable product), the minimum MTBF, and the average MTBF.

In a failure-truncated plan, the MTBF is computed at the occurrence of predetermined number of failures. If this MTBF (operating hours divided by number of failures) is higher than the value suggested by the handbook, the lot is accepted.

In time-censored plans, the handbook suggests the number of units to be tested for a predetermined number of hours. It also gives the number of failures permitted.

A sequential test plan requires little testing at one time. If a product is not acceptable or rejectable, then more units are tested. This plan is the most popular. The main advantage of this plan is that a decision can be reached very quickly if the product is very good or very bad.

Before a sequential test, the customer specifies producer's risk, consumer's risk, and discrimination ratio (ratio of average MTBF and minimum acceptable MTBF). From these values, a chart such as that in Fig. 3.18 is selected. A customer may also specify minimum test duration and minimum number of pieces to be tested. If such conditions are not specified, it is up to the supplier to test any number of pieces (as few as one). At each failure, the cumulative failures are plotted against cumulative test hours (in multiples of minimum acceptable MTBF). If the point of their intersection crosses the *accept* lines, the entire lot is accepted. If this point crosses the *reject* lines, the lot is rejected. If the point is in the *continue test* area, the tests are continued.

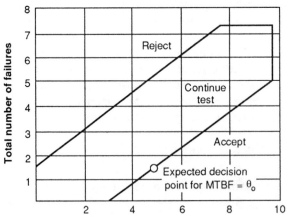

Figure 3.18 A reliability demonstration plan for Mil-Std-781.

3.4.4 Burn-in and screening

Burn-in and screening tests are 100 percent production tests designed to catch infant mortality failures. If a component is manufactured properly, it should not require these tests, and the aim should be to eliminate them by removing the causes of the problems. But, since new causes can crop up, audits should always be conducted. In burn-in tests, devices are subjected to a constant high temperature. Screening tests use a profile of operating environments such as temperature cycling, vibration, and humidity. Screening may require special tests other than environmental tests. For example, leakage current was monitored in the tests illustrated in Fig. 3.19 to identify weak components that might have not been noticed otherwise.

Many companies are forced into doing these tests because the supplier does not. These tests are also done at circuit board level and assembly level to reveal poor workmanship (such as defective solder joints) or loose connections. These tests also show whether the design can withstand vibrations and temperature cycling.

The type of stress and the length of the test are generally suggested by the appropriate military standard, which is based on experience. Unfortunately, many practitioners take these suggestions as the final word; as a result, screening may not be long enough and weak products may get out to the customer. The manufacturer should first find the stresses that will shake out all weak components and second make

INFANT MORTALITY VERSUS DESIGN FAILURES

Figure 3.19 Weak components identified by functional screening.

sure the test duration extends to the point where failure rate is constant.

To reduce test time, one can use accelerated test conditions to flush out weak failures faster. The only requirement is that the failure mode on accelerated and unaccelerated tests should be the same. Some use combined environmental reliability tests (CERT) for screening. These tests combine temperature cycles, vibration, humidity, dust, and other environments with the equipment powered. Weak failures have been known to appear faster. Automotive products, for example, see these environments together in use.

Testing philosophy. The nature of screening tests, which are often called *environmental screening*, depends on the use profile of the system. A test may include vibration, temperature cycling, burn-in, run-in, humidity, and special stresses unique to the system use. Most electronics manufacturers screen 100 percent of the product. A wise strategy, before embarking on a test program, is to first determine which tests are necessary, according to the use profile, and which components must be tested. If a good manufacturing process is in place, sample testing instead of 100 percent testing will be adequate insurance.

There should also be a rational analysis of whether the test should be done at component level, board level, or system level. The following cost information, derived from electronics industry rules of thumb can aid in making the right choice:

Test level	Approximate cost of finding and replacing a defective component
Component	$ 5.00
Board	$ 50.00
Black box	$ 500.00
System	$2500.00 and up

These numbers show that, if system reliability is low, testing should be done at the black-box level. If the black-box reliability is low, testing should be done at the circuit board level. Not all the boards need be 100 percent tested—only the questionable ones. Similarly, only the questionable components need to be tested 100 percent. Conversely, if reliability at system level is very high, system-level testing may be unnecessary. The same test philosophy applies at black-box level and below.

Testing at board level is generally desirable, even though a few manufacturers trust their suppliers enough to ignore board-level screening. But, if there are 100 components on a board, each having 99 percent reliability, the board reliability before soldering (assuming the components are mounted properly) is 36.6 percent. This means that only one out of three boards can be expected to work satisfactorily, and this is not an uncommon situation. The reliability is further reduced by such elements as insertion of wrong components, reversed polarities, and poor soldering. If one out of four boards contains some assembly defect, the board yield is shaved by another 25 percent. For such situations, screening at board level is cost-effective.

Role of automated testing. If the chance of finding a reliable board is roughly 1 in 3 before soldering, it pays to do screening at the component level because, as the preceding table shows, it is 10 times cheaper ($5 versus $50) to replace a bad component at that level. Functional testing at the component level is rarely justified, however, unless it is done on automated test equipment (ATE). Whether testing is automated or not, the prudent course is to first determine whether the supplier is already doing the screening. Some components are sensitive to environments such as electrostatic discharges and fail in spite of the screening done by the supplier. If these "dead on arrival" (DOA) components are numerous, rescreening may be desirable.

The investment in intelligent, programmable automated test equipment is justified when:

- The volume of components is large.

- There are numerous tests for components and boards. The parallel test capabilities of the equipment can drastically cut test time.

- Reliability can be improved by diagnostic testing.

- The component location and the cause of its failure must be identified quickly.

Automated testing permits effective screening at the component level because the test time is short enough to fit into a production schedule. If screening is not done at this level, much more time must be spent at board level. Also, at board level, the repair process can cause failure of other components.

Whether testing is done at board level or component level, the total cost of inspection, troubleshooting, test, and repair should be kept as low as possible. If a board contains about 200 components and is produced at a rate of about 100,000 a year, the author's experience shows that automated test equipment can pay for itself within 2 years. The savings come primarily from a reduction in test and repair time.

ATE offers many other advantages. It can test a component alone or on a circuit board. When a complete board with 200 or so components is tested, the ATE can isolate defective components. A major benefit to designers and quality control engineers is that ATE summarizes vast amounts of data on design parameters such as leakage currents and various voltages—valuable information for determining statistical distributions for vendors' products.

Dynamic burn-in. Dynamic burn-in requires functional testing during burn-in. Software built into automated test equipment can control the functional tests. Dynamic burn-in greatly improves the field reliability of the boards because they have, in effect, gone through a system test. Boards tested this way tend to be less than 2 to 10 percent defective as shipped. A simple system of five printed-circuit boards showed the following reliabilities after standard and dynamic burn-in:

Test reliability	Outgoing board reliability, %	System reliability, %
Standard burn-in	91.41	63.82
Dynamic burn-in	98.55	92.96

Economies from choosing the right tests. Most manufacturers do not use the correct tests; they are happy when they do not find failures. Absence of failure is not an indication of a good circuit board. Burn-in at low temperature may help weed out very weak failure modes, but the product leaving the door may still be unreliable. Performing a good test on a statistically selected sample is a good strategy. A good test forces the unreliable components to fail. It may require 100 percent testing initially and may reduce the manufacturing yield at first, but it improves the yield for the customer. Later, manufacturing yield can be improved and 100 percent testing can be eliminated with a good failure analysis and management program. The savings accomplished in this way can be enormous, and customer satisfaction can be expected to soar.

Caution: The correct test will be an accelerated test based on use profile. Effective screening can take months if the test is not accelerated.

Example 3.6 A computer display oscillated frequently in the humid summer environment only. This is an indication that humidity testing is one of the right tests.

Economies from tradeoffs. Some manufacturers do not perform environmental screening; they test circuit boards only functionally. Weak components pass the quality control tests but fail later, in the field. These manufacturers do not understand that a tradeoff can be made between the cost of screening and the increase in reliability. An increase in reliability means lower warranty costs for the manufacturer and lower repair and maintenance costs for the customer.

One method of evaluating a tradeoff is to compute life-cycle costs (LCC) for alternatives. The following is a simplified example of such a model:

LCC = initial cost + screening cost + warranty failures cost

+ repair and maintenance costs to customer

+ potential loss of business because of poor reliability

For a medium-size business, the following costs were calculated for a printed-circuit board having a service life of 5 years:

Market price: $2500

Screening cost: $15

Warranty cost without screening: $45.95

Warranty cost with screening: $27.57

Repair and maintenance cost (charged to customer): $95.40

Repair and maintenance cost (with screening): $66.78

Potential loss of business (because of low reliability): $56.25

Potential loss of business if board is screened but still not competitive in design: $33.75

LCC without screening = $2500 + $0 + $45.95 + $95.40 + 56.25

= $2697.60

LCC with screening = $2500 + $15.00 + $27.57 + $66.78 + $33.75

= $2643.10

The savings per board ($2697.60 − $2643.10 = $54.50) may not appear significant, for 200,000 boards, the total savings are $54.50 × 200,000 = $10,900,000. Of this savings, 47.5 percent went to the manufacturer and the remainder went to the customers. Such a situation, in which both the supplier and customers save, is a major benefit of assuring quality.

3.5   Reliability in the Use Phase

The usual tests are unable to catch all failure modes, many of which are discovered in the field. Too many of these failures indicate short-comings in the reliability effort in the earlier stages. A closed-loop failure management system should not only solve the problem permanently but should also improve the design review process and the design of the tests.

The biggest challenge facing quality producers is competition on longer warranties. When warranties are involved, low reliability can easily destroy a business. At the time of writing, the following warranties were in effect:

Automotive: 3-year warranty on all parts; 7 years or 70,000 miles on power train and corrosion

Water meters: 25 years warranty on any malfunction or failure

Power transformers: 15 years

Radio-frequency tags: 100 years; 150 years under consideration

These warranties are standard features of the products. The manufacturers do not charge their customers extra for them, in the hope of getting more business. The reliability effort in the use phase therefore must be aimed at detecting trends as early as possible and making design changes expediently. A good tool for establishing trends early is Weibull analysis. A Japanese company gets information on worldwide failures immediately through communication satellites and performs Weibull analysis weekly. With Weibull analysis, warranty costs can be predicted easily.

Trends like this are spreading globally. The best practice is to track the distribution of field failures. If the distribution starts during the warranty period, the failure modes should be thoroughly analyzed and removed permanently.

3.6   Topics for Student Projects and Theses

1   Explore why industry and universities use an exponential model for reliability when practically no system has demonstrated exponential probability. (*Note:* This is the author's opinion drawn from experience of 25 years. There are opposing views, but readers should make their own conclusions.)

2   Most popular standard sampling plans for reliability are based on exponential distribution (for example, Mil-Std-781, Mil-Hdbk-H108). If real distributions in the field are not exponential (see Wong[17]), should industry still use the standard plans? Support your argument.

3 Does the bathtub profile seem to apply to every product? Are there products or systems where it does not apply? What kinds of distribution would you expect in the first region, infant mortality?

4 Can you combine the three regions of the bathtub profile into one model? If so, what model would you propose?

5 Each distribution on a Weibull graph has a unique slope. Will two normal distributions with identical means but different standard deviations be represented by the same line? If so, how can you differentiate the two distributions on the same graph?

6 Why does industry use MTBF for mechanical products when the failure rate is known to be not constant? Can you use MTBF if the failure rate is not constant?

References

1. [New computer introduced by Coleco Industries in 1985 (in the United States)]
2. M. B. Darch, Chap. 3, in *Reliability and Maintainability of Electronic Systems*, J. E. Arsenault and J. A. Roberts (eds.), Computer Science Press, Potomac, Md., 1980.
3. *USAF R&M 2000 Process*, Handbook SAF/AQ, U.S. Air Force, Washington, D.C., 1987.
4. *Reliability Prediction of Electronic Equipment*, Mil-Hdbk-217.
5. *Specification for Capacitor, Fixed, Tentalum*. Mil-C-39003.
6. *Handbook of Reliability Prediction Procedures for Mechanical Equipment*, Eagle Technology for Logistics Division, David Taylor Research Center, Carderock, Md., and Product Assurance and Engineering Directorate, Belvoir Research Development and Engineering Center, Fort Belvoir, Va.
7. *Procedures for Performing a Failure Mode, Effects, and Criticality Analysis*, Mil-Std-1629.
8. D. Raheja, "Failure Mode and Effects Analysis—Uses and Misuses," *ASQC Quality Congress Transactions*, 1981.
9. James T. Duane, "...," *IEEE Transactions on Aerospace*, April 1964.
10. "...," Mil-Hdbk-189,
11. C. R. Lipson, and N. Sheth, *Statistical Design and Analysis of Engineering Experiments*, McGraw-Hill, New York, 1979.
12. Patrick D. T. O'Connor, *Practical Reliability Engineering*, Wiley, New York, 1985.
13. *Low Cycle Fatigue*, Special Technical Publication 942, ASTM, Philadelphia, 1988.
14. *Standard General Requirements for Electronics Equipment*, Mil-Std-45, Naval Forms and Publications Center, 5801 Tabor Avenue, Philadelphia, PA 19120, 1984.
15. *Sampling Procedures and Tables for Life and Reliability Testing*, Mil-Hdbk-H108 Naval Forms and Publications Center, Philadelphia, 1960.
16. *Reliability Testing for Engineering Development, Qualification, and Production*, Mil-Std-781,
17. Kamon Wong, "The Roller-Coaster Curve Is In," *Quality and Reliability International*, January-March 1989.

Further Reading

Arsenault, J. E., and J. A. Roberts, *Reliability and Maintainability of Electronic Systems*, Computer Science Press, Potomac, Md., 1980.

Bazovsky, I., *Reliability Theory and Practice*, Prentice-Hall, Englewood Cliffs, N.J., 1961.

Carter, A. D. S., *Mechanical Reliability*, Macmillan, New York, 1986.

Fuqua, Norman B., *Reliability Engineering for Electronic Design*, Marcel Dekker, New York, 1987.

Ireson, Grant W., and Clyde F. Coombs, *Handbook of Reliability Engineering and Management*, McGraw-Hill, New York, 1988.

Jensen Finn, Petersen, *Burn-In*, John Wiley and Sons, New York.

Johnson, L. G., *Theory and Technique of Variation Research*, Elsevier, Amsterdam, 1964.

Johnson, L. G.: *The Statistical Treatment of Fatigue Experiments*, Elsevier, Amsterdam, 1964.

Kapur, K. C., and L. R. Lamberson, *Reliability in Engineering Design*, John Wiley and Sons, New York, 1977.

Lewis, Elmer E., *Introduction to Reliability Engineering*, John Wiley and Sons, New York, 1987.

Lipson, C., and N. J. Sheth, *Statistical Design and Analysis of Engineering Experiments*, McGraw-Hill, New York, 1979.

Definition of Terms for Reliability and Maintainability, Mil-Std-721,

Reliability Design Qualification and Production Acceptance Tests, Mil-Std-781,

Reliability Program for Systems and Equipment Development and Production, Mil-Std-785,

Reliability Assurance Program, Mil-Std-790,

Murphy, Ray W., *Endurance Testing of Heavy Duty Vehicles*, Publication SP-506, SAE, Warrandale, Pa., 1982.

Nelson, Wayne, *Applied Life Data Analysis*, Wiley-Interscience, New York, 1982.

Automated Electronics Reliability Handbook, Publication AE-9, SAE, Warrandale, Pa, 1987.

Shooman, M.: *Probabalistic Reliability: An Engineering Approach*, McGraw-Hill, New York, 1968.

Singpurwalla, Nozer D.: "Inference from Accelerated Life Tests and When Observations Are Obtained from Censored Samples," *Technometrics*, February 1971, pp. 161–170.

Chapter

4

Maintainability Engineering

4.1 Maintainability Engineering Principles

Reliability deals with reducing the frequency of breakdowns. *Maintainability* deals with the duration of breakdowns. Many practitioners as well as academicians think maintainability refers to easy repair and maintenance. But the aim of maintainability is to prevent the occurrence of downtime and prevent the need for maintenance altogether. Only when this is not feasible (because of limited-life components) does the goal become to minimize downtime.

A report released by the U.S. Government Accounting Office in 1989 revealed dismal statistics on maintenance costs. The government paid large sums for maintenance which was supposed to be covered by warranties. During a 12-month warranty period, the Navy reported 251 failures on antimissile systems, but the contractor paid none of the repair costs. The warranty stated that the company was responsible for repairs only if the Navy experienced more than 5238 failures. In 1987, the Army reported that it paid $9.9 million in warranties for engines on its M-1 tanks, but was reimbursed for only $10,453 worth of claims. The contractor refused to pay many claims because they were not filed within the required 90 days. Such examples abound not only in government but also in private industry. One of the big reasons is lack of organization for tracking warranty agreements.

These examples show that the customer often pays twice for corrective maintenance, at least during the warranty period. The cost of the associated downtime is huge and unfortunately does not even get quantified. Sooner or later, these costs become a big burden. The solution here is to increase investment in maintainability, not in maintenance.

Maintainability is mainly a design function; most of the effort should be expended in the design phase. It should be part of every design review. One cannot emphasize enough that maintainability has tremendous influence on maintenance and other support costs, which are usually the largest segment of life-cycle costs.

Maintainability, according to Mil-Hdbk-472,[1] is the ability of an item to be retained in or restored to specific conditions when maintenance action is performed by personnel having specified skill levels and using prescribed procedures and resources at each prescribed level of maintenance and repair.

This definition does not emphasize design enough. As a result, many design engineers have diverted their attention from minimizing maintenance requirements, often to the point where maintenance becomes extremely difficult. In a large manufacturing machine, for example, it was necessary to inspect the condition of a gear periodically. The preventive maintenance procedure took 8 to 16 hours because the technician had to open about 30 bolts, then call an electrician (as required by union rules) to remove electrical connections, then remove several parts before obtaining a good view of the gear. With maintainability analysis, the design was improved to eliminate all such extra work. The new equipment had a camera installed inside, and the gear was visible much more clearly on a television screen. Downtime due to routine inspection: zero hours!

With this example in mind, it is time to redefine maintainability. The definition this author has used for years is:

> Maintainability is the science of minimizing the need for maintenance and minimizing the downtime if maintenance action is necessary. It is the probability that downtime (excluding logistics delays) will be below a specified duration.

This definition emphasizes the aggressive role of the designer in reducing the operating and support costs. It is up to the reader to decide which definition to choose. In either case, the maintainability objectives can be met.

Maintainability itself is not quantified, since it depends on qualitative parameters "using prescribed procedures and resources." The argument is that rarely are the prescribed procedures adequate, and the resources such as spares and trained personnel are not available all the time. Three basic indexes of maintainability are

- Equipment repair time
- Inherent availability
- Mean downtime for maintenance

Equipment repair time (ERT) is the time an average technician will take to repair an item. It is the time from preparing for repair to testing to make sure the equipment is working right. Mil-Hdbk-472 contains the detailed listing of tasks. Equipment repair time does not include waiting time and logistic delays.

ERT is probabilistic in nature. Some technicians will repair quickly, others will be slow. If the majority of the technicians are highly trained or not trained enough, then the distribution is skewed. One can use the median value or the following rules of thumb:[2]

If the time-to-repair distribution is symmetrical and bell-shaped, the typical ERT is equal to the arithmetic average:

$$ERT = \text{mean time to repair} = MTTR$$

If the time-to-repair distribution is exponential:

$$ERT = 0.69\,MTTR$$

If the time to repair has a log-normal distribution, as is usually the case:

$$ERT = 0.45\,MTTR$$

The second measure of maintainability, *inherent availability*, actually is a measure of maintainability and reliability together. It is the uptime ratio designed into the system, and can be defined as

$$A_i = \frac{\text{uptime}}{\text{total scheduled time}} = \frac{\text{uptime}}{\text{uptime} + \text{active downtime}}$$

Most defense contracts assume that uptime is proportional to MTBF and active downtime is proportional to MTTR. The inherent availability then becomes

$$A_i = \frac{MTBF}{MTBF + MTTR}$$

This definition, even though used industrywide, has several flaws. MTBF, the mean time from one failure to another, may include logistic delays as well as downtime. Generally, the records for computing MTBF are not adequate.

The definition of inherent availability is the ideal case. It assumes the spares are available all the time and repair is done immediately. It does not include waiting or logistic delay time because the design engineer has no way of predicting them, unless some complex mathematical modeling is done. The design engineer can predict MTTR on the assumption that the spares and the trained personnel are available. There is, however, another side to the coin. The customer is concerned about actual availability, called *operational availability*, in which waiting time and delays are a rule rather than exception. A design engineer must also try to maximize operational availability if possible. Operational availability can be defined as

$$A_o = \frac{\text{uptime}}{\text{uptime} + \text{active repair time} + \text{mean logistic downtime}}$$

$$= \frac{\text{MTBF}}{\text{MTBF} + \text{MTTR} + \text{MLDT}}$$

Quantifying the terms in the A_o equation requires a planned data collection effort that very few are willing to mount. If one wants to know the actual availability, then accurate logs of uptime or downtime must be kept. Some companies have running clock-hour logs built into their equipment. They can then compute operational availability, which may be redefined as

$$A_o = \frac{\text{uptime}}{\text{total scheduled time}} = \frac{\text{scheduled time} - \text{downtime}}{\text{scheduled time}}$$

The third measure of maintainability, *mean downtime for maintenance* (MDT), is the mean downtime spent on corrective as well as preventive maintenance:

$$\text{MDT} = \frac{F_c M_{ct} + F_p M_{pt}}{F_c + F_p}$$

where F_c = frequency of corrective maintenance
F_p = frequency of preventive maintenance
M_{ct} = average time for corrective maintenance
M_{pt} = average time for preventive maintenance

4.2   Maintainability during the Design Phase

4.2.1   Developing maintainability specifications

Typical maintainability requirements such as equipment repair time, inherent availability, and mean downtime are not sufficient for specifying maintainability for complex equipment, which today contains complex software and microprocessors. There are many false fault alarms—as much as 50 percent false in some cases. No longer can a technician isolate a fault without very expensive automated test equipment. Therefore, fault diagnosis and fault isolation capability must be designed into the product. This will permit maintenance to be performed quickly.

If the equipment is designed with backup features, maintenance can be performed during the off shifts, without inconvenience to users.

Such equipment forgives failures from many causes, but requires more maintenance on backup components.

Mil-Hdbk-338[2] suggests MTTR computed from the log normal distribution (if applicable) as the primary measurement for maintainability. In addition the following indexes may be used:

- Maximum corrective maintenance time at a specified percentile, such as the 90 or 95 percentile.

- Percent faults that can be isolated to a single replaceable item.

- Mean maintenance worker-hours per repair.

- MTBM—mean time between maintenance.

- MTBM(C)—mean time between corrective maintenance (corrective maintenance may not be related to a component failure).

- MTBM(P)—Mean time between preventive maintenance.

- MTBA—mean time between adjustments.

- %NFF—percent no faults found (applies, for example, to a circuit board that is found to be functional but did not work in a system because of a temporary fault. It may not respond at system level because of vibration, shock, etc.). Sometimes these faults are as much as 50 percent of total faults.

- %CND—percent faults that cannot be duplicated at the repair facility.

- %BCS—percent faults bench check serviceable. These are faults checkable at the depot.

- Testability index—(1) percent digital "stuck-at" zero or 1 faults testable for microprocessors; (2) a quality rating of the ability to make testing easier for any equipment. A comprehensive checklist for a testability rating system is in Sec. 4.2.9.

- Mean downtime for software errors.

- %DTS—percent downtime due to software errors.

- %DTH—percent downtime due to hardware failures.

- %DTS—percent downtime from system failures (both hardware and software faults count as failures).

- Safe life for safety-critical component replacement.

- Safe life for calibration of safety devices.

- Safe life for damage tolerance (crack growth).

It is advisable that goals for these items be included in the specification as soon as possible. On the basis of these goals, the maintainability program plan is developed; this is the master list of tasks and milestones to be monitored during the entire life cycle.

4.2.2 Design review for maintainability

Maintainability review starts in preliminary design. Before a design is approved, the maintainability features should be established. Especially to be considered are

How much downtime is acceptable? (Specify availability.)

At what level will the repair be made? At component level? Circuit board level? Black box level?

Who will do the repair? The contractor? The customer? A technician? An engineer? Or should the failed product be thrown away? (The answers will affect the levels of skills required.)

Does the equipment have to be designed for inspectability?

Should the equipment be designed for ease of accessibility?

Can any built-in test features be added to reduce downtime?

Is modular design helpful? For hardware? For software?

Is fault-diagnostic software helpful?

Is there a need to isolate the faults?

What kinds of skills are required by the repair technicians?

What kinds of inspection and test methods are required for crack detection (aircraft, nuclear plants)?

What are the provisions for replacing safety-related components?

What are the allowable crack sizes?

What are the corrosion-prevention measures?

What kind of maintainability tests are required?

Answers to such questions as these determine the design concepts for maintainability. All questions are aimed at reducing or eliminating downtime.

It is very important at this stage to study the weaknesses and the strengths of competitors. A dramatic example shows why: A very famous North American maker of business machines lost its entire 90 percent market share of desktop copiers to Canon, a Japanese company. The Canon copier did not require preventive maintenance, while the North American copier required a $1200 per year maintenance contract. Besides, the Canon copier sold for only $1100 in 1981 while the North American manufacturer charged $3000.

The author bought the Canon copier and used it for 7 years without any downtime or preventive maintenance, except for changing the ink cartridge (which took less than a minute). The copier was still in excellent condition at the time of this writing. That's maintainability: a practically maintenance-free design! Granted that the author's sample size was only one, the fact that Canon increased its market share from 9 to 90 percent indicates that other customers had similar experience.

If competitors' equipment requires much maintenance and many spares, maintainability by design is a newcomer's opportunity to offer a better value. The manufacturer that thrives on selling spare parts will make profits in the short term but market death is almost certain in the long run, as the fate of the formerly dominant desktop copier manufacturer testifies.

4.2.3 Maintainability analysis

As design progresses, it becomes important to analyze such requirements as human factors, hierarchy of repair, diagnostics, fault isolation, and testability. For example, consider the level of repair. There are at least seven options:

- Repair at component level
- Repair at subassembly level (example: circuit board level)
- Repair at assembly or black box level
- Replace at subassembly level and repair later
- Throw away the subassembly
- Throw away the assembly
- Throw away the entire product

Maintainability analysis may include decisions such as preventive maintenance hours required per operating hour and the cost of downtime. The right way to approach the design is: First, decide how you can do better than the competition and be able to market maintainability as a selling feature. Second, evaluate the life-cycle costs for each option. If the life-cycle costs are too complex to estimate, one can compare the option on some meaningful measure such as cost per circuit mile for a power company or cost per million instructions for a computer network. The final decision should be such that it benefits both the supplier and the customer.

4.2.4 FMECA for maintainability

One of the best-known tools for maintainability analyses is failure-mode, effects, and criticality analysis. Although FMECA was originally applied to reliability analysis, it can easily be tailored to maintainability by assessing the criticality of a failure in terms of downtime costs. The goal is to minimize the risk of unavailability. The desirable corrective action therefore is to modify the design so it requires the least amount of downtime.

Figure 4.1 shows a portion of a maintainability FMECA for a large electromechanical system. As an example, consider the failure modes for pipe in this figure. Two possible causes of failure are corrosion and cracks in the solder at the joint. The best way to prevent downtime from these causes is to eliminate them. If this is not feasible, one can design for ease of maintenance. Failure from corrosion can be eliminated by changing the pipe material from copper to stainless steel. Failure at the solder joint can be eliminated by substituting a flexible joint for the stiff one; a flexible joint can withstand much larger structural loads.

Unfortunately, most design engineers do not perform FMECA for maintainability. Even if they do, their corrective action usually is to perform more preventive maintenance instead of improving the design. This is usually costly to the user. If users have to pay for the high cost of downtime, they can take their business to a better supplier.

One reason maintainability has not received proper recognition is that project managers readily see the increased design cost it incurs but fail to see the benefits. Increases in customer satisfaction and in market share are rarely perceived as major components of the life-cycle cost model. There is yet another roadblock. The engineer is heavily criticized for increasing product design costs and rarely credited for the tenfold reduction in support costs that maintainability brings.

Remember that any product that can advertise reduced downtime is a highly desirable product. Most customers are willing to pay extra if they are convinced it means less maintenance; Sony televisions, Honda Accord cars, and IBM 3061 computers attest to this fact.

4.2.5 Maintainability prediction

Maintainability prediction is done for the indexes chosen in the specification. As an example, consider the prediction of inherent availability,

$$A_i = \frac{\text{MTBF}}{\text{MTBF} + \text{MTTR}}$$

This requires prediction of two parameters: MTBF and MTTR. An equipment MTBF and MTTR prediction will depend on what components are likely to fail. The values will be based primarily on infor-

Part name/part number	Potential failure modes	Causes (failure mechanism)	Effects	Expected downtime	Frequency per year	Recommended improvement	Maintenance requirements
Pipe	Leakage in pipe	1. Corrosion	Loss of Freon	8 h	2	Use stainless steel pipe	None
		2. Temperature cycling	Loss of Freon	8 h	5	Monitor temperature with thermocouples	Replace probes every 120 days
	Leakage at the joint	1. Cumulative fatigue	Loss of Freon	14 h	4	Monitor vibration with accelerometers	Calibrate accelerometers semi-annually
		2. Poor soldering	Loss of Freon	2 h	2	Provide flexible plaster coupling at the joint	None
Valve	Sticky, intermittent	1. Dirt or foreign objects	Loss of control of temperature	5 h	2	Electronic redundant valve action and fault identification	Clean every 15 days
	Stuck open	1. Component wearout	Loss of control of temperature	4 h	1	Electronic redundant valve action and fault identification	Check every 3 months
	Stuck closed	1. Component failed	Loss of control of temperature	4 h	2	Electronic redundant valve action and fault identification	Check every 3 months
		2. Component expansion and contraction	Loss of control of temperature	5 h	3	Electronic redundant valve action and fault identification	None

Figure 4.1   An example of FMECA for maintainability.

mation from the FMECA. The MTBF is computed as the reciprocal of the total failure rate. The MTTR is computed as the average of expected repair times for various breakdowns; an industrial engineer can roughly estimate the hours required for the repair, and Mil-Hdbk-472 gives 26 generic steps required for repairing electronic equipment (Fig. 4.2). For new technology the FMECA can help by pointing out the additional failure modes.

Predictions of other appropriate indexes suggested in Sec. 4.2.1 can be made from historical data. The predictions pinpoint the opportunities for improving the design.

4.2.6   Life-cycle cost analysis

During maintainability analysis, life-cycle costs are used frequently to assess the savings in support costs for various options. Maintainability usually has the highest influence on downtime costs and therefore the LCC. In general, life-cycle costs are costs to society. Specifically, they are costs to (1) the supplier, (2) the user, and (3) innocent bystanders who might get killed or injured or whose property might be damaged from such mishaps as aircraft accidents. As far as maintainability is concerned, bystanders also include operating personnel and maintenance and repair personnel. One can write the LCC model as

$$LCC = \text{nonrecurring costs} + \text{recurring costs}$$

Examples of nonrecurring costs are

- Basic engineering
- Test and evaluation
- Program management and control
- Manufacturing and quality engineering
- Nonrecurring production costs
- Experimental tooling

Examples of recurring costs are

- Spares
- Repair labor cost
- Inventory entry and supply management
- Support equipment
- Technical data and documentation
- Training and training equipment

Category	Elemental activity	Activity no.
Preparation	System turn-on, warm-up; setting dials and counters as necessary.	1
	Activity 1 plus time awaiting particular component stabilization.	2
	Opening and closing radome.	3
	Gaining access and reinstalling covers (other than radome).	4
	Obtaining test equipment and/or tech orders.	5
	Checking maintenance records.	6
	Procuring components in anticipation of need.	7
	Setting up test equipment.	8
Malfunction verification	Observing indications only.	1
	Using test equipment to verify malfunctions inherently not reproducible on ground.	2
	Performing standard test problems or checks.	3
	Testing for pressure leaks.	4
	Attempting to observe elusive or nonexistent symptom(s).	5
	Using special test equipment designed specifically for this equipment.	6
	Making a visual integrity check.	7
Fault location	Fault self-evident from symptom observation.	1
	Interpreting symptoms by mental analysis only (from knowledge/experience).	2
	Interpreting displays at different settings of controls.	3
	Interpreting meter readings.	4
	Removing unit(s)/subunit(s) and checking in shop.	5
	Switching and/or substituting unit(s)/subunit(s).	6
	Switching and/or substituting part(s).	7
	Removing and checking parts.	8
	Making a visual integrity check.	9
	Checking voltages, continuity, waveforms, and/or signal tracing.	10
	Consulting tech orders.	11
	Conferring with tech reps or other maintenance personnel.	12
	Performing standard test problem(s).	13
	Isolating pressure leak.	14
	Using special test equipment designed specifically for this equipment.	15

Figure 4.2 FMECA for maintainability.

Category	Elemental activity	Activity no.
Part procurement	Obtaining replacement component from aircraft spares or tool box.	1
	Obtaining replacement(s) from bench, shop, or preissue stock.	2
	Obtaining replacement component(s) by cannibalization.	3
	Attempting to obtain replacement component(s). Unavailable.	4
Repair	Replacing unit(s)/subunit(s).	1
	Replacing parts.	2
	Correcting improper installation or defective plug-in connection(s).	3
	Making adjustments in aircraft.	4
	Making adjustments in shop.	5
	Baking magnetron.	6
	Precautionary repair activity (includes so-called fault location, part procurement, and repair times spent when symptom not verified).	7
	Repairing wiring or connections.	8
Final malfunction test	Function checkout following completion of repair.	1

Figure 4.2 (*Continued*)

- Logistics management
- Operators
- Utilities

Other versions of the model are

LCC = (initial price + warranty costs to the company)
+ (repair, maintenance, and operating costs to the customer)

and

LCC = acquisition cost + cost of operation, repair, and maintenance
+ cost of downtime + cost of liabilities + cost of lost sales

The last model is most appropriate, but many companies find it very difficult to quantify the cost of downtime and the cost of lost sales. They use the other two models, which reveal only part of the facts.

The real world has many hidden costs. They are usually unaccounted for but are huge. They include costs of excess spare parts inventory, idle manpower, missed schedules, promises not kept to customers, time spent in meetings and solving problems, engineering changes, loss of goodwill, and penalties.

These costs are not easy to estimate. At the same time, ignoring them will lead to a very unrealistic model. Therefore, a reasonable estimate based on a consensus should be included. Other cost factors, if they are related and significant, should also be included. Examples: increases in interest rates, fuel costs, power costs, and labor costs.

One of the best uses of a life-cycle cost model is to evaluate the merits of several options. Assume two options for a desktop copier. Design A is a traditional design which requires periodic replacement of the toner (black ink powder, cost: $50 per year). The parts must be cleaned by a maintenance technician, and limited-life components must be replaced when they fail. A technician is needed to do the repairs. A maintenance contract for the periodnot under warranty costs $1200 per year. The selling price is $2500. The warranty cost, paid by the manufacturer, is $750.

Design B has a sealed cartridge containing ink and copying components. The cartridge has to be replaced twice a year to ensure a fresh supply of ink; it can easily be replaced by the user in less than a minute. The cartridge itself is reliable, and costs $75. Since most of the technology is in the cartridge, there are few other critical components, and they are designed not to fail during the expected 5-year life of the copier. Just in case an external component should break down, the lightweight copier can be hand-carried to the dealer, who can fix it in minutes.

By using a simple LCC model, the two options can be compared. Assume that design B costs $2800, somewhat more than design A because of new technology. Assume also that design A, like design B, has a 5-year life.

Design A:

LCC = (manufacturer's cost) + (maintenance costs and

downtime costs to customer)

= (selling price + warranty costs during first year) +

(cost of maintenance for 4 years + cost of toner +

cost of downtime)

= ($2500 + $750) + ($1200 × 4 years + $200 + $1100*)

= $9350

Design B:

LCC = (selling price + warranty costs at 5% of selling price) +

(expected cost to repair for 4 years + cost of cartridges)

= ($2800 + $140) + ($600 + $75 × 10)

= $4290

The conclusion is in favor of design B even if its initial cost is higher. The distribution of costs makes design B even more attractive to the customer:

	Supplier cost, $	Buyer cost, $
Design A	3250	6100
Design B	2940	1350

The supplier will save $310 per unit with design B ($3250 − $2940), but the buyer will benefit far more, saving $4750 ($6100 − $1350). If the gain in market share that is likely for design B is taken into account, the unit cost of design B will be far lower.

4.2.7 Design for accessibility

Accessibility means having sufficient room around a component to diagnose, troubleshoot, and complete maintenance work in a safe and

---

*Estimated cost of downtime over 5 years.

effective manner. Provision must be made for movement of necessary tools and equipment with consideration for various body positions.[3]

Detailed plastic scale models can be used in assessing accessibility. If these are not possible, then the volume available around the equipment for activities must be estimated from the drawings. The former approach is much more effective, since it provides an opportunity to test for accessibility and ease of maintenance.

4.2.8 Design for ease of maintenance

Ease of maintenance means making activities at the human/ equipment interface easier. It can be assured in many ways.

Minimize maintenance in the first place. In automobiles, the fuel injection method of ignition has eliminated the need to check the condition of the distributor.

Allow repairs with least handling. A person changing a component on an automobile should not be required to remove a major assembly.

Design for reliability of the interface of mating parts. If an integrated circuit (chip) on a circuit board is going to be removed and replaced a few times, it should not be soldered to the circuit board. It should be mounted on a gold socket (aluminum sockets may not provide a good contact after about three replacements).

Design for off-line repair. A repair on a very large bearing in an automated process plant was expected to result in 2 days' downtime. However, the material handling equipment was designed so that the bearing could be removed and repaired off-line. A spare bearing was installed to reduce the process downtime to less than 1 hour.

Provide fault tolerance. This approach makes use of an alternate means already installed so that the downtime is practically zero. The failed primary unit can be repaired while the secondary line is substituted. Fault-tolerant equipment often permits preventive maintenance while operation continues. The IBM 3081 mainframe computer, for example, performs self-maintenance without the customer being aware of it.

Plan tools and equipment in advance. Pillar[3] suggests useful rules of thumb: "If a service device or a tool is required to construct a piece of equipment, it will also be needed later to repair it.... When considering features for ease of maintenance major emphasis should go to the weight of components that must be moved." This means additional equipment such as a forklift truck or a crane may be required.

Design for remove-and-replace instead of repair.   It may be cheaper to replace the whole circuit board than to have a technician isolate the faulty component.

Use automated test equipment.   The automated test equipment (ATE) can be a portable intelligent system which can generate thousands of test patterns to isolate a faulty path.

Design for testability.   This design approach has become very important for reducing the cost of troubleshooting. With microprocessors doing most of the work, the equipment can be designed to self-test and make highly skilled technicians unnecessary for troubleshooting complex electronics. Technicians with less skills can make the repair in minutes instead of days. Microprocessors built into the product can be programmed to test the product and isolate the fault. Some are designed to heal themselves switching over to an alternate means or a backup device.

Testable designs incorporate two features. First, the microprocessor itself must be capable of being tested to assure its own performance. Second, the product (in which the microprocessor resides) should also lend itself to ease of testing involving many features other than the microprocessors.

Microprocessors with as many as 240,000 gates and millions of combinations of paths require millions of test patterns for testing. With physical limits on the number of input and output pins, it is almost impossible to perform all the tests. Some paths cannot be tested at all unless appropriate test points are monitored. Figure 4.3 shows a simple circuit with only two gates. In this circuit there are two input pins, x_1 and x_2, and one output pin z. Columns x_1 and x_2 in the figure show all the possible inputs. The output for each combination of inputs is shown in column z. Columns A/0, B/0, C/0, D/0, E/0, and F/0 show the output if either of the points A through F is faulty, being stuck at zero. The remaining columns show the output if either of the points is faulty, being stuck at digital one.

It is not easy to isolate the faults. If all the inputs at x_1 and x_2 produce the outputs in column z, one may think the circuit is functioning correctly. This, however, is not necessarily a correct conclusion; fault condition C/1 also gives the same output for the identical inputs. Under these conditions, the test will indicate a good circuit.

Columns A/0, B/0, E/0, D/1, and F/1 produce identical outputs. This means that these faults cannot be isolated in the circuit test if only $x_1, x_2,$

Complexity in testing and fault isolation

All single stuck faults

x_1	x_2	z	A/0	B/0	C/0	D/0	E/0	F/0	A/1	B/1	C/1	D/1	E/1	F/1
0	0	1	1	1	1	1	1	0	0	0	1	1	1	1
0	1	1	1	1	1	1	1	0	1	0	1	1	1	1
1	0	0	1	1	0	0	1	0	0	0	0	1	0	1
1	1	1	1	1	0	0	1	0	1	1	1	1	0	1

Figure 4.3 A simple circuit with only two gates.

and z are monitored. From a maintainability viewpoint, however, technicians should be provided with the means to distinguish and isolate important faults. If these paths are known to be safety-critical during the design stage, the design engineer is often able to design for testability. Recently several books have been published on this subject.[4,5]

Modular designs. This approach requires designs to be divided into physical and functionally distinct units to facilitate removal and replacement. It enables subassemblies and assemblies to be designed as removable and replaceable units for an enhanced design with minimum downtime. These units, called modules, can be a black box, a circuit board, or a major assembly. A few advantages of modularity are[6]

1. New designs can be simplified and design time can be shortened by making use of standard, previously developed building blocks.
2. Only low-level technician skills are required.
3. Training is easier.
4. Engineering changes can be done faster, with fewer side effects.

Design for disposal at failure. This approach is modular also, except that the failed unit is not repaired. This can sometimes be costly when many electronic devices are thrown away because of temporary faults. In one automotive company, as much as 40 percent of returned circuit boards were found to have no fault. On the other hand, a few advantages of this approach are

1. Savings in repair time, tools, facilities and manpower
2. Improved reliability because of reduced maintenance-induced errors
3. Improved interchangeability because of standardization

Additional disadvantages are

1. Increased inventory cost because more replacement parts must be stocked, since repairs cannot be made
2. Costs of engineering changes are higher because of larger inventory

Design for working environment. Where possible, the environmental factors affecting the ability of personnel to perform work should be considered. Environments such as temperature, radiation, humidity, chemicals, dust, and inadequate illumination increase the chance of maintenance-induced errors. Figure 4.4, for example, shows the effect of temperature on the average number of mistakes.

Standardization. Components such as nuts, bolts, lockwashers, and fuses can be easily standardized. This makes the right part much easier to obtain and cheaper. Inventory costs are reduced drastically, and reliability is improved. In one organization more than 360 part numbers were reduced to only 45 part numbers by standardization.

Design for interchangeability. A product can be designed for functional or physical interchangeability. In the former case, two or more items serve the same function. In the latter case, the same fastening, connecting, and mounting means can be used for more than one item.

Caution: Watch out for safety concerns. Interchangeability can result in accidents. See Chap. 5.

Design for technician transparency. The U.S. Air Force program for the R&M 2000 process[7] requires a revolutionary approach: standard operating, maintenance, and testing characteristics across similar subsystems. The same technician can then work on all subsystems. This technician "transparency" will reduce manpower requirements, because specialists for certain subsystems are not needed. Any techni-

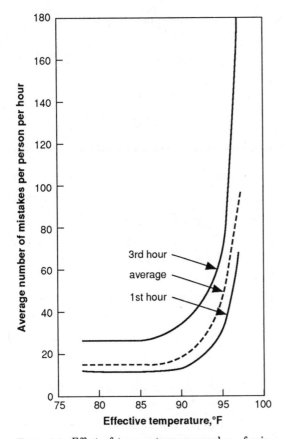

Figure 4.4   Effect of temperature on number of mistakes. (*Source: AMC Pamphlet AMCP 706-134, Headquarters, U.S. Army Materiel Command, pp. 10–12.*)

cian can isolate a fault to the lowest field-replaceable module. It also reduces the skill level required by the technicians.

4.2.9   Design for ease of testing

There are many strategies which can make testing easier and more efficient—that can enhance *testability*. Mil-Std-2165 includes a large list of strategies and scoring guidelines to assess design effectiveness for testability.[8] These strategies are in the following checklist.

Mechanical design

Is a standard grid layout used on boards to facilitate identification of components?

Is enough spacing provided between components to allow for clips and test probes?

Are all components oriented in the same direction (pin 1 always in same position)?

Are standard connector pin positions used for power, ground, clock, test, and other signals?

Are the number of input and output (I/O) pins in an edge connector or cable connector compatible with the I/O capabilities of the selected test equipment?

Is defeatable keying used on each board to reduce the number of unique interface adapters required?

When possible, are power and ground included in the I/O connector or test connector?

Have test and repair requirements affected decisions on conformal coating?

Is the design free of special setup requirements (e.g., special cooling) which would slow testing?

Test access

Are unused connector pins used to provide additional internal node data to the tester?

Are signal lines and test points designed to drive the capacitive loading represented by the test equipment?

Are test points provided that let the tester monitor and synchronize with on-board clock circuits?

Are test access points placed at those nodes that have high fan-out?

Are buffers employed when the test point is a latch and susceptible to reflections?

Are buffers or divider circuits employed to protect those test points that may be damaged by an inadvertent short circuit?

Are active components, such as multiplexers and shift registers, used to make necessary internal node test data available to the tester over available output pins?

Are all high voltages scaled down within the item before the test access point so as to be consistent with tester capabilities?

Is the measurement accuracy of the test equipment adequate for the tolerances established for the item being tested?

Parts selection

Is the number of different part types the minimum possible?

Have parts been selected which are well-characterized in terms of failure modes?

Are the parts independent of refresh requirements? If not, are dynamic devices supported by sufficient clocking during testing?

Is a single logic family being used? If not, is a common signal level used for interconnections?

Analog design

Is one test point per discrete active stage brought out to the connector?

Is each test point adequately buffered or isolated from the main signal path?

Are multiple interactive adjustments prohibited for production items?

Are functional circuits (amplifiers, regulators, etc.) of low complexity?

Are circuits functionally complete, without bias networks or loads on another unit?

Is the minimum number of multiple phase-related or timing-related stimuli required?

Is the minimum number of phase or timing measurements required?

Is the minimum number of complex modulation or unique timing patterns required?

Are stimulus frequencies compatible with tester capabilities?

Are stimulus rise time or pulse width requirements compatible with tester capabilities?

Do response measurements involve frequencies compatible with tester capabilities?

Are response rise time or pulse width measurements compatible with tester capabilities?

Are stimulus amplitude requirements within the capability of the test equipment?

Are response amplitude measurements within the capability of the test equipment?

Does the design avoid external feedback loops?

Does the design avoid or compensate for temperature-sensitive components?

Does the design allow testing without heat sinks?

Are standard types of connectors used?

Digital design

Does the design contain only synchronous logic?

Are all clocks of differing phases and frequencies derived from a single master clock?

Are all memory elements clocked by a derivative of the master clock? (Avoid having elements clocked by data from other elements.)

Does the design avoid resistance-capacitance one-shot circuits and dependence on logic delays to generate timing pulses?

Does the design support testing of "bit slices"?

Does the design include data wraparound circuitry at major interfaces?

Do all buses have a default value when unselected?

For multilayer boards, is each major bus laid out so that current probes or other devices can be used for fault isolation beyond the node?

Is a known output defined for every word in a read-only memory (ROM)? Will the improper selection of an unused address result in a well-defined error state?

Is the number of fan-outs for each internal circuit limited to a predetermined value?

Are latches provided at the inputs to a board in those cases where tester input skew could be a problem?

Is the design free of wired ORs?

Does the design include current limiters to prevent domino-effect failures?

If the design incorporates a structured testability design technique (e.g., scan path, signature analysis), are all the design rules satisfied?

Are sockets provided for microprocessors and other complex components?

Built-in testing

Can built-in testing (BIT) in each item be exercised under control of the test equipment?

Is the test program set designed to take advantage of BIT capabilities?

Are on-board BIT indicators used for important functions? Are BIT indicators designed so that a BIT failure will give a FAIL indication?

Does the BIT use a building-block approach (e.g., are all inputs to a function verified before that function is tested)?

Does building-block BIT make maximum use of mission circuitry?

Is BIT optimally allocated in hardware, software, and firmware?

Does on-board ROM contain self-test routines?

Does BIT include a method of saving on-line test data for the analysis of intermittent failures and operational failures that are nonrepeatable in the maintenance environment?

Is the failure rate contribution of the BIT circuitry within stated constraints?

Is the additional weight attributed to BIT within stated constraints?

Is the additional volume attributed to BIT within stated constraints?

Is the additional power consumption attributed to BIT within stated constraints?

Does the allocation of BIT capability to each item reflect the relative failure rate of the items and the criticality of the items' functions?

Are BIT threshold values, which may require changing as a result of operational experience, incorporated in software or easily modified firmware?

Is processing or filtering of BIT sensor data performed to minimize BIT false alarms?

Are the data provided by BIT tailored to the differing needs of the system operator and the system maintainer?

Is sufficient memory allocated for confidence tests and diagnostic software?

Does mission software include sufficient hardware-error-detection capability?

Is the failure latency associated with a particular implementation of BIT consistent with the criticality of the function being monitored?

Are BIT threshold limits for each parameter determined as a result of considering each parameter's distribution statistics, the BIT measurement error, and the optimum fault detection versus false alarm characteristics?

Test requirements

Has level-of-repair analysis been accomplished? Should the replacement be done at component level or circuit board level, or at assembly level?

For each maintenance level, has a decision been made for each item on how built-in testing, automatic test equipment, and general-purpose electronic test equipment will support fault detection and isolation?

Is the planned degree of test automation consistent with the capabilities of the maintenance technicians?

For each item, does the planned degree of testability design support the decisions on the level of repair, test mix, and degree of automation?

Are the test tolerances established for BIT consistent with those established for higher-level maintenance tests?

Test data

Do state diagrams for sequential circuits identify invalid sequences and indeterminate outputs?

If a computer-aided design system is used for design, does the CAD data base effectively support the test generation process and test evaluation process?

For large-scale integrated circuits used in the design, are data available to accurately model the circuits and generate high-confidence tests for them?

For computer-assisted test generation, is the available software sufficient in terms of program capacity, fault modeling, component libraries, and postprocessing of test response data?

Are testability features included by the system designer documented in terms of purpose and rationale for the benefit of the test designer?

Is a mechanism available to coordinate configuration changes with test personnel in a timely manner?

Are test diagrams included for each major test? Is the diagram limited to a small number of sheets? Are intersheet connections clearly marked?

Is the tolerance band known for each signal on the item?

4.3 Maintainability in the Manufacturing Stage

This section covers minimizing the downtime on manufacturing equipment. Companies such as M&M Mars, Firestone, General Motors, Intel, Caterpillar, and Eastman Chemicals (Kodak) have applied maintainability techniques to save millions of dollars. One company estimated that a 2 percent reduction in downtime will result in a savings of $36 million over 5 years on an annual sales volume of $800 million.

4.3.1 Maintainability for existing equipment

When the equipment has already been procured, major design changes are usually out of the question. However, many modifications can still be made. Sometimes life-cycle cost analysis will show that it may be cheaper to scrap the equipment and buy new.

The first step is to identify the 20 percent of problems that are responsible for 80 percent of the downtime (a famous principle). The next step is to analyze the causes of downtime by a systematic analysis such as fault-tree analysis. The author uses process analysis maps (PAMs), which are simplified fault trees. The details are covered in Chap. 6. Any maintenance technician can be trained to construct PAMs. In fact, a company should train all maintenance technicians to use such tools. Technicians can use analysis tools effectively and come up with many useful ideas—but, of course, management must listen in order to benefit from them.

Example 4.1   A manufacturing process used two rollers, jacketed in 1-inch-thick rubber (Fig. 4.5) to squeeze chemicals from thin, wet sheets of steel at high speed. The rubber jacket blew into pieces roughly once a week, like a tire blowing out on a semi truck. It tore the steel sheets apart, resulting not only in very poor quality control, but also in a severe safety problem. One worker had his neck lacerated. The total downtime cost to recover from failure was roughly $250,000 per week.

The engineers tried to improve the design by substituting better materials for the rubber jacket. They did not make significant progress. They tried better bonding compounds between the steel and the jacket, again with no progress. They also tried making rough surfaces and grooves on the steel so that the rubber would be held more tightly. Two years went by and $24 million were surrendered to downtime.

(a)

(b)

Figure 4.5   The chemical rinse process. (*a*) End and cross section views of rollers. (*b*) Partial process analysis map.

solution   A systematic analysis solved the problem in a few hours! The process analysis map in Fig. 4.5 was constructed with the help of the maintenance technicians. Five major causes were identified. The technicians assessed the percent contribution of each cause. The figure shows that the highest contributor to the failure was heat buildup in the steel roller (70 percent contribution), which was unable to transfer heat except through the rubber jacket. Since the technicians were trained to eliminate the root cause of the downtime, one of them took no time in coming up with the solution. The following questions and answers led to the solution:

Q. What is the most probable cause of downtime?

A. Excess heat buildup in the steel roller.

Q. If the cause of downtime is excess heat, can we eliminate excess heat?

A. Yes. We can circulate refrigerated drinking water inside the steel roller. That should remove a lot of heat. I can bring a pump from the stockroom and circulate water from the drinking water fountain.

Q. Excellent idea. Why haven't you tried?

A. I have been telling management about this for 5 years, but no one cares.

Maintainability is a gold mine—if someone is willing to mine. At the author's recommendation, the technician was allowed to use his idea. It worked. After the change, the author never saw a single incidence of downtime in the 2 years of his association with the company! Savings: $12 million per year.

4.3.2 Maintainability for new equipment

Maintainability for new manufacturing equipment is the same as maintainability for a new product, except that the supplier may team up with the customer in analyzing maintainability if it is a custom design. For off-the-shelf equipment, maintainability information may not be available to the user. The author has worked with several suppliers of large capital equipment items priced in millions of dollars and found that, although the approach is still in infancy, the two most effective tools are failure-mode, effects, and criticality analysis (FMECA), covered in Chap. 3, and fault-tree analysis (FTA), covered in Chap. 5. Other valuable techniques are quality function deployment, design reviews, and maintenance engineering safety analysis (Chap. 5).

Example 4.2   A pattern-recognition optical system for manufacturing very large scale integrated circuits was to be procured by a large semiconductor manufacturer. A fault tree was constructed by a team of engineers (Fig. 4.6). An independent team performed FMECA. (If the same team performs FTA and FMECA, the results are likely to be the same.) A sample sheet of the FMECA is shown in Fig. 4.7. Both teams came up with different problems, with few exceptions. By the time the analysis was complete, over 100 design concerns were identified. At least 50 design changes were determined to be important. The time of about eight engineers for 3 days was a small investment compared to the value of 50 design improvements. Since the system was still on the drawing board, the cost of engineering changes was insignificant.

4.4   Maintainability in the Test Stage

Maintainability tests are often called *maintainability demonstration tests*. At least four types of tests can be conducted:

1. Test for inherent equipment downtime.

2. Test for human variation in performing the maintenance work.

3. Maintenance level tests. These are done to verify the ability to repair at organization, intermediate, and depot level.

4. Tests for maintenance during emergencies.

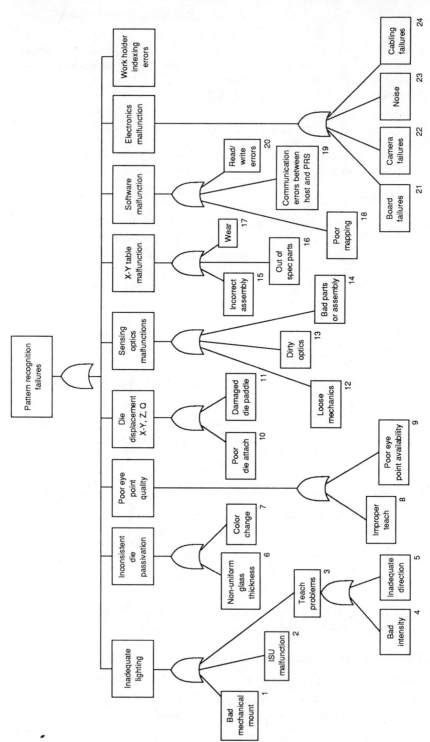

Figure 4.6 Fault-tree analysis for pattern-recognition equipment.

Failure mode	Causes	Effects	Criticality	Design action	Fault verification	RCM action
Optics malfunction	Ambient heat	Permanent deformation	II A	Provide fan	Warn of fan failure	Check fan tolerances every 2 months
	Dirt	Erroneous output	II B	Add filter	Not required	Replace filter monthly
Circuit parameters drift	High leakage current	Parameters out of control	II D	Qualify critical components	Not required	Install software to monitor parameters
	Dirt on circuit	Intermittent performance	II B	Conformal coat	Not required	Not required
	High junction temperature	Degraded performance	II B	Derate parts below 50%	Not required	Use infrared inspection
X–Y table innaccurate	Supplier design	False output	II A	Perform FMEA with supplier	To be determined	To be determined
	Horizontal position drift	False output	II A	Software control	Not required	Check eccentricity during routine maintenance

Figure 4.7 FMECA for pattern-recognition equipment.

Maintainability tests start during product development. The tests can be done with the help of portions of the system, mockups, and models. In some cases, a detailed analysis may be the only option. For example, the analyses may suggest that a forklift truck is required. If the access area is too small, this is the best time to redesign the access area. It is too expensive to make changes later.

4.4.1 Prerequisites for maintainability tests

Before expensive tests are done, some strategy is needed. The tests should not be based only on guesswork. No maintainability test for a complex equipment should be performed without an FMECA. This analysis reveals many important failure modes—those that are likely to result in downtime. Ignoring this analysis will result in many design changes at a huge cost later.

The other pertinent pieces of information required may be maintainability specifications, where all required parameters are specified (if the specification does not contain maintainability information, then it should be developed); past histories on similar equipment so that such problems are removed from design; and interviews with repair and maintenance crews.

4.4.2 Tests for inherent equipment downtime

Demonstrations of inherent downtime may consist of tests for testability, fault diagnosis, fault isolation, interchangeability, simulation, and any automated repair such as switching over to a backup device.

Two types of verification may be done: (1) to verify if the downtime for each failure mode is within specified limits and (2) to assure that the allocation of maintainability goals is adequate.

4.4.3 Tests for human variations

Human variation tests are statistical in nature. A probability distribution is constructed to determine the downtime distribution. Tests are conducted for such parameters as mean time to repair (MTTR), mean downtime (MDT), mean worker-hours per operating hour (MWH/OH), and maximum time to repair at a certain percentile.

Several technicians should test for each task to construct the distribution of downtime. The variation in repair times for mechanical systems is very large. Besides, this information will help in workforce planning. If it would be too expensive to test all the tasks, then they may be selected. The author follows the strategy of selecting tasks related to safety and those which can result in high downtime. Usually repair times are distributed in a log-normal manner. Chapter 2 contains an example of the construction of this distribution.

4.4.4 Maintenance level tests

These tests are conducted to see if the maintenance tasks designed to be done at the operational level (organizational level) can indeed be performed by the operations technician. Similar verification can be done for the ability to perform maintenance and repair tasks at the intermediate level (sometimes called *unit level*), and at the depot level (in an automotive environment, depots are equivalent to *remanufacturing centers*).

4.5 Maintainability in the Use Stage

Maintainability analysis is a continuous improvement process. Many causes of downtime are not known until the system is in the field. There are at least three major areas where costs can be reduced through design improvements:

1. Prediction and reduction of limited-life items

2. Monitoring and predicting operational availability

3. Minimizing maintenance and support costs through maintainability engineering

4.5.1 Prediction and reduction of limited-life items

Limited-life items are those that require repair or replacement during the operating life of the system. Several texts contain models based on the assumption that the failure profile of the limited-life item is a normal distribution or an exponential distribution. If these assumptions are not justified, which is usually the case, a better tool is Weibull analysis. This reliability analysis can give insight to the number of spares required by the system. It includes exponential as well as the normal distribution cases. This distribution applies to several families of data. It uses actual data (even partial data will help) in making an assessment of the distribution. Chapters 2 and 3 cover several examples of this analysis. An example is shown here to demonstrate the application of this technique for spares prediction.

Example 4.3 A sample of 25 bearings out of a lot of 400 was tracked for time to failure in a brewery. These bearings are used in similar equipment systemwide.

At the time of analysis, only 8 bearings had failed. Their time to failure in equipment hours is converted to weeks below. How many spares are required between weeks 30 and 40?

Bearing serial number	Weeks to failure
1	6.0
2	11.0
3	2.2
4	7.8
5	12.2
6	4.3
7	6.2
8	8.4

solution   The data are ranked below and plotted against median rank (see Chap. 2) as the y coordinate:

Weeks to failure	Median rank value
2.2	2.73
4.3	6.62
6.0	10.55
6.2	14.49
7.8	18.44
8.4	22.38
11.0	26.32
12.2	30.27

The growth for the data is shown in Fig. 4.8. The Weibull line is projected from the plot of the first eight points, based on a sample size of 25.

From the graph:

Percent expected to fail in 30 weeks = 89.5 percent

Percent expected to fail in 40 weeks = 97.0 percent

Percent expected to fail during the period 30 to 40 weeks = 97.0 – 89.5 = 7.5 percent

A conclusion can be made that 7.5 percent of the bearings from the 400 procured can be expected to fail during the period between 30 and 40 weeks. Therefore, 400 × .075 = 30 bearings are expected to fail during this period.

4.5.2 Monitoring and predicting operational availability

Availability is the ultimate objective of maintainability. Tracking data are important information that leads to design improvements. One method of obtaining such data is to keep records on downtime.

Figure 4.8 Weibull graph for bearing failure, in weeks.

Then availability is

$$A = \frac{\text{uptime}}{\text{total scheduled time}} = \frac{\text{total scheduled time} - \text{downtime}}{\text{total scheduled time}}$$

If there is significant amount of downtime, the causes may be analyzed. Often, minor design changes will yield as much as a 100-fold reduction in costs.

Another method of assessing availability is to use Monte Carlo simulation. The steps and an example are given below:

1. Establish the model for availability.
2. Construct statistical distributions for each element of the model.
3. Randomly draw a value of each element from its probability distribution.
4. Compute availability from the data obtained in step 3.
5. Construct the distribution of availability.

The simulation can be done manually, but a computer is normally used to handle the large amount of data.

Example 4.4

Step 1:   Consider this model for availability:

$$A = \frac{\text{MTBF}}{\text{MTBF} + \text{MDT}}$$

Step 2:   Construct Weibull plots for MTBF and MDT from actual data.

Step 3:   Randomly pick a value of MTBF from the Weibull plot. Also, randomly pick a value of MDT from the Weibull plot.

Step 4:   Compute availability using the equation in step 1.

Step 5:   Repeat the first four steps several hundred times using a computer program. This will result in hundreds of values for the expected availability. Construct the probability distribution of availability (see Chap. 2). The distribution will give minimum and maximum values of availability. It will also give the probability associated with each availability target.

4.5.3 Minimizing support costs

Minimizing support costs is an important goal, but customer satisfaction is just as important. The two goals go together.

Support costs consist of costs of an ongoing maintainability effort, preventive maintenance, corrective maintenance, and many other logistics costs. An investment in maintainability is required to reduce the cost of preventive and corrective maintenance.

The investment should continue as long as the total of the investment and the support cost is lower. The difference between the initial total cost and the total cost in any year is the direct savings. Figure 4.9a shows year-by-year performance of total cost. It shows that the minimum cost is achieved at the sixth year. Should the investment in maintainability be stopped? The answer is no, because the customer service goal, as shown in Fig. 4.9b, has not been achieved (the downtime goal). Some may argue that extra cost in achieving customer ser-

(a)

(b)

Figure 4.9   Tradeoff between maintainability investment and service. (a) Year-by-year cost performance with minmum cost goal achieved in sixth year. (b) Minimum cost goal is achieved but the service goal is not achieved.

vice is not justified; the total cost will no longer be the minimum. Actually, the cost will be lower when the customer is satisfied. A satisfied customer brings more customers and more business volume. This should lower the unit cost of the product and boost profits.

4.6 Topics for Student Projects and Theses

1 Suggest the best definition of maintainability as far as you are concerned. Suggest measures or models.

2 Write a maintainability specification for a product.

3 Write a maintainability demonstration plan for a product.

4 Describe your approaches to designing maintenance-free products.

5 Conduct a study to assess the impact of maintainability on life-cycle costs for a product.

6 Describe how fault-tree analysis can be used for maintainability analysis.

7 Write a generic procedure for a design review for maintainability.

8 Write your approach to testing for maintainability.

References

1. *Maintainability Prediction*, Mil-Hdbk-472, Naval Forms and Publications Center, 5801 Tabor Avenue, Philadelphia, PA 19120, 1984.
2. *Electronic Reliability Design Handbook*, Mil-Hdbk-338, Naval Forms and Publications Center, 5801 Tabor Avenue, Philadelphia, PA 19120, 1988.
3. Charles S. Pillar, "Maintainability in Power Plant Design," *Proceedings of the Sixth Reliability Engineering Conference for Electric Power Industry*, American Society for Quality Control, Milwaukee, 1979.
4. Hideo Fujiwara, *Logic Testing and Design for Testability*, MIT Press, Cambridge, Mass., 1985.
5. Parag K. Lala, *Fault Tolerance and Fault Testable Designs*, Prentice-Hall, Englewood Cliffs, N.J., 1985.
6. *Engineering Design Handbook*, Pamphlet AMCP 706-134, Headquarters, U.S. Army Materiel Command, 1970.
7. *R&M 2000 Process*, U.S. Air Force,
8. Mil-Std-2165.

Further Reading

Arsenault, J. E., and J. A. Roberts, *Reliability and Maintainability of Electronic Systems*, Computer Science Press, Potomac, Md., 1980.
Blanchard, B. S., and E. E. Lowery, *Maintainability Engineering, Principles and Practices*, McGraw-Hill, New York, 1969.

Goldman, A. S., and T. B. Slattery, *Maintainability, A Major Element of System Effectiveness,*

Mil-Std-470, *Maintainability Program Requirements*

Mil-Std-471, *Maintainability Verification/Demonstration/Evaluation*

Patton, Joseph D., Jr.: *Maintainability and Maintenance Management,* Instrument Society of America, N.C., 1980.

5

System Safety Engineering

5.1 System Safety Principles

System safety is defined as the application of engineering and management principles, criteria, and techniques to optimize safety within the constraints of operational effectiveness, time, and cost throughout all phases of the system life cycle.[1] This definition suggests that merely a good design is not the goal; rather, the goal is a successful design, one that anticipates unknown problems in advance, is implemented on time, and minimizes life-cycle costs.

Often tests are not done under conditions that duplicate field environments, and many human errors, such as miswiring in an aircraft, are easy to make. The safety profession, which deals with preventing loss of human lives, cannot wait for reports on accidents. By that time too many may have died! A safety professional may accept the probability of loss of human life to be less than one per million per year. Since no scientist or engineer can predict such a low probability with any data, attempts are made to eliminate the product hazards altogether. This concern for preventing accidents without the help of much data is responsible for expanding the concept of product safety to system safety.

To understand system safety, one must understand the definition of a *system* as "...a composite, at any level of complexity, of personnel, procedures, materials, tools, equipment, facilities, and software. The elements of this composite entity are used together in the operational or support environment to perform a given task or achieve a specific mission."[1]

This definition has a special meaning in system safety. It implies that it is the mission safety that is important. It also implies that all the hardware in the system, including the interfacing hardware, since it can cause an accident, must be compatible. Similarly, all the internal and external software that affects mission safety must be compatible. One must look beyond one's own product to assure safety of the mission. This concept applies even to simple products such as toys. If a certain toy has small parts that can come loose, then the toy design is unsafe; the manufacturer cannot rely on parents to prevent a child from swallowing the parts. Since it is not easy to predict the behavior of parents, the goal should be to design a toy whose parts cannot be swallowed, even if the parents are totally careless. System safety uses such an approach.

To assure mission objectives, the system boundaries must be defined. The boundaries for system safety are subjective. They include interfaces among hardware, software, environments, and humans. The extent to which these factors influence mission safety will determine the boundaries on systems. For example, an electronic system in an automobile would include a mechanical system in its boundary if the mechanical parts degrade the performance of the electronics. Consider the early warning system on an aircraft called *engine condition monitoring* (ECM). The mission of this equipment is to report on the health of the jet engines before they become sick. System safety deals with predicting as many lifetime hazards as possible and controlling them by sound designs. The following may be considered as boundaries for ECM.

The hardware associated with the mission, such as:

- Sensors that may measure incorrect exhaust temperature, compressor pressure, engine speed, fuel consumption, and quietness of operation

- A computer that may malfunction during data analysis

- A printer that prints out wrong reports for maintenance technicians

- Display equipment in the cockpit that may display misleading information

Software associated with the mission, such as:

- Analytical software in the ECM computer that may have logic faults

- Software for cockpit displays that may have coding errors

- Fault isolation software that may give wrong indication

Humans in the system, such as:

- A pilot who may misunderstand warnings
- Technicians who may not follow procedures
- Software engineers who may introduce errors while modifying the software
- Maintenance personnel whose actions may give wrong warnings
- Manufacturing personnel who may not install the sensors right, may miswire, or may mount components backward

Procedures in the system, such as:

- Preventive maintenance procedures that may be ambiguous
- Configuration control procedures that may not get implemented
- Software change procedures that may be lax

Materials in the system, such as:

- Fuel for engines
- Lubricants
- Deicing agents
- Chemicals used during cleaning, maintenance, and repair
- Deodorants used in passenger cabins

Environments of the system, such as:

- High altitude, which may cause erroneous readings in some measurements
- Hot-and-cold cycles that may degrade the accuracy and precision of instruments
- A large number of landings, which may cause cracks
- Vibration, which can cause loose connections

Looking at the partial list above, it is easy to see that there are many roadblocks to system performance and hence safety. Not to be forgotten is peripheral equipment, which may introduce additional hazards. Unless hazards in all the areas above are controlled by design, it is impossible to feel confident about safety.

Example 5.1 Consider a baby who has a heart defect. The parents want to monitor the breathing of the baby so they can immediately take corrective action when needed. The objective here is to monitor breathing performance 24 hours a day. Ordinarily the parents would consider as the boundary an item of medical equipment that can sound an alarm. If this equipment is truly the boundary, then it should be enough to accomplish the mission safely, and the safety analysis should be limited to it. But the fact is that even if the equipment works reliably, the mission can be unsafe. The system boundary contains some more factors: the parents may not be able to hear the alarm, or the alarm may not be sounded because of electromagnetic interference from a videocassette recorder or because the remote controller of a TV may interfere with the execution of software in the equipment. The boundary is now extended to the analysis of the parental patterns, the electronic equipment used in the house, and the surroundings which may accidentally cut off the equipment. Even the baby may accidentally disconnect the equipment from the power source. If the parents are going to be outside the house, a remote alarm system should also be included in the system boundary.

5.1.1 System safety process

The system safety process is a risk management process developed to the highest degree. The steps in the process are

1. Identify the risks using hazard analysis techniques as early as possible in the system life cycle.

2. Develop options to eliminate, control, or avoid the hazards.

3. Provide for timely resolution of hazards.

4. Implement the best strategy.

5. Control the hazards through a closed-loop system.

System safety is not only a function of engineering but is an integral part of top management activities. Participation of management can assure the timely identification and the resolution of hazards. Therefore, a major requirement of system safety is that it must be institutionalized. The management tasks as well as the engineering tasks are well identified in Mil-Std-882. They are listed in Table 5.1.

5.1.2 Risk assessment

Risk is a function of the severity of an accident and its probability of occurrence. Typically, various types of analyses will produce hundreds of hazards for a complex equipment. No organization can afford to react to all the hazards. The hazards are assigned priorities so that at least the catastrophic and critical accidents are prevented. The standard severity categories in Mil-Std-882 are

Catastrophic (I)   Death or system loss

Critical (II)   Severe injury, severe occupational illness, or minor system damage

TABLE 5.1   Management and Engineering Tasks according to Mil-Std-882

			Applicability*			
Task	Title	Type†	Conceptual	Validation	Full-scale engineering development	Production
100	System safety program	Mgt	G	G	G	G
101	System safety program plan	Mgt	G	G	G	G
102	Integration/management of associate contractors, subcontractors, and AE firms	Mgt	S	S	S	S
103	System safety program reviews	Mgt	S	S	S	S
104	System safety group/system safety working group support	Mgt	G	G	G	G
105	Hazard tracking and risk resolution	Mgt	S	G	G	G
106	Test and evaluation safety	Mgt	G	G	G	G
107	System safety progress summary	Mgt	G	G	G	G
108	Qualifications of key system safety personnel	Mgt	S	S	S	S
201	Preliminary hazard list	Eng	G	S	S	N/A
202	Preliminary hazard analysis	Eng	G	G	G	GC
203	Subsystem hazard analysis	Eng	N/A	G	G	GC
204	System hazard analysis	Eng	N/A	G	G	GC
205	Operating and support hazard analysis	Eng	S	G	G	GC
206	Occupational health hazard assessment	Eng	G	G	G	GC
207	Safety verification	Eng	S	G	G	S
208	Training	Mgt	N/A	S	S	S

TABLE 5.1   Management and Engineering Tasks according to Mil-Std-882 (*Continued*)

Task	Title	Type†	Applicability*			
			Concep-tual	Valida-tion	Full-scale engi-neering devel-opment	Produc-tion
209	Safety assessment	Mgt	S	S	S	S
210	Safety compliance assessment	Mgt	S	S	S	S
211	Safety review of ECPs and waivers	Mgt	N/A	G	G	G
212	Software hazard analysis	Eng	S	G	G	GC
213	Government fur-nished equipment/ governemnt fur-nished property system safety anal-ysis	Eng	S	G	G	G

*Applicability codes: S—selectively applicable; G—generally applicable; GC—generally applicable to design changes only; N/A—not applicable.
†Task type: Eng—system safety engineering; Mgt—management.

Marginal (III)	Minor injury, minor occupational illness, or minor system damage
Negligible (IV)	Less than minor injury, occupational illness, or system damage

The probability classifications are listed in Table 5.2.

TABLE 5.2   Probability Classifications from Mil-Std-882

Description	Level	Frequency of occurrence	
		Individual items	Fleet or inventory
Frequent	A	Likely to occur frequently	Continuously experienced
Probable	B	Will occur several times in life of an item	Will occur frequently
Occasional	C	Likely to occur sometime in life of an item	Will occur several times
Remote	D	Unlikely but possible to oc-cur in life of an item	Unlikely but can reasonably be expected to occur
Improbable	E	So unlikely, it can be as-sumed occurrence may not be experienced	Unlikely to occur, but possi-ble

TABLE 5.3 Hazard Priority Guidelines from Mil-Std-882

Hazard risk index	Suggested criteria
IA, IB, IC, IIA, IIB, IIIA	Hazard unacceptable
ID, IIC, IID, IIIB, IIIC	Hazard undesirable (higher management decision is required)
IE, IIE, IIID, IIIE, IVA, IVB	Acceptable with review by management
IVC, IVD, IVE	Acceptable without review

To determine which hazards should receive priority, a system of making decisions should be established. Suggested guidelines from Mil-Std-882, based on combinations of severity and frequency, are in Table 5.3. This approach offers a good structure for risk control. In practice, however, it is difficult to implement. Management participation in resolution of unacceptable hazards has been very meager.

5.1.3 Technical risk analysis

Since severity categories I and II deal with catastrophic and major accidents, they usually comprise over 80 percent of the risk. This does not, however, mean category III and IV hazards should be neglected, but much less attention is required. In deciding whether to accept the risk from unresolved hazards, an organization should develop a model for assessing their impact. The most practical model, which can be translated into dollars, is

$$\text{Risk} = F \times S \qquad \text{summed for all category I and II hazards}$$

where F is the frequency of the hazard and S is the severity of the hazard in dollars. The severity of a hazard in dollars is computed from the value of the personnel liability and property damage, as estimated for that hazard.

If one wishes to assess total technological risk, the risk model must contain not only the hazards but also the failures not related to safety. This can be written as

$$\text{Technological risk} = F \times S \qquad \substack{\text{summed for all critical} \\ \text{and major failures}}$$

whether or not the failures result in an accident.

Typical publications quantify risk as a product of severity and probability or frequency, but they fail to state a period for the probability or frequency. Is it 1 year, 2 years, or the lifetime? They also fail to clarify when to use probability and when to use frequency. The word *probability* refers to one accident, but in the lifetime of a system there can be multiple accidents. Hence, *frequency* should be used if the risk is considered over the entire life cycle. The relation between frequency and probability is

$$F = PNn$$

where P = probability of an accident during a period
  N = number of equipment items
  n = number of periods over the intended life

5.1.4 Residual risks

If some hazards are uncontrolled and management cannot or does not want to control them, then someone must assume the risk. At least the following options are available:

1. Avoid risk by not marketing the product.

2. Transfer risk to a contractor or an insurance company and pay for it in advance.

3. Accept the risk and keep doing what comes naturally. This is the most undesirable approach but also the most common.

4. Prevent risk by switching to another technology or different procedure.

5. Control risk by eliminating the hazards of high priority.

5.1.5 Safety margins

Safety verification and validation are tools to assure safety. The validation process consists of making sure that either all the identified hazards have been resolved or their risk has been accepted. The verification process involves testing of critical features for safety margins. This involves destructive and nondestructive testing to verify susceptibility to sudden failure.

5.1.6 Emergency preparedness

Management must ensure that safety is inherent and there are no operational hazards. This can be verified by conducting emergency drills, covered in Chap. 8, and by institutionalizing a closed-loop reporting and correction system for all hazards. That means that operating personnel must report all "incidents" which have not but could result in accidents.

5.2 System Safety in Design

Since system safety engineering is the process of applying scientific and engineering principles, criteria, and techniques, its goal is to develop products that are immune to component failures and human errors. To develop robust designs, it is important to understand generic causes of accidents. The definitions given below will help.

Safety: Freedom from those conditions that can cause death, injury, occupational illness, or damage to or loss of equipment or property.

Hazard: A condition which can result in a mishap or accident under certain conditions.

Hazardous event: An occurrence that creates a hazard.

Hazard severity: An assessment of the worst credible mishap that could be caused by a specific hazard.

Mishap: An unplanned event that results in death, injury, occupational illness, or damage to equipment or property.

Accident: A hazard triggered into an unsafe condition which may result in a mishap.

Risk: An expression of the possibility of a mishap in terms of hazard severity and hazard probability (or frequency).

System safety management: An element of management that defines the system safety program requirements and ensures the planning, implementation, and accomplishment of system safety tasks and activities consistent with the overall program requirements.

System safety program plan: A description of the planned methods to implement the tailored requirements of a standard (Mil-Std-882 or equivalent), including organizational responsibilities, resources, methods of accomplishment, milestones, depth of effort, and integration with other engineering and management activities.

The definition of a hazard above indicates that it is only a condition that can result in a mishap. For a hazard to result in mishap, something else also must happen. This is called a *trigger event*. If either the hazard or the trigger event is prevented, the mishap can be prevented, as Fig. 5.1 shows. Examples of hazards and triggers are

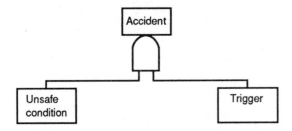

Figure 5.1   Generic causes of an accident.

Hazard	Trigger event
High temperature	Temperature is above flash point
Nuclear gas leakage	Operator does not notice
Train driver is under the influence of drugs	Train on wrong track close to an oncoming train
Cracks in aircraft fuselage body	Too many landings and takeoffs

5.2.1 Criteria for a safe design

A safe design is one that accomplishes at least the following four goals:

1. *It minimizes the gravity of danger in normal use:*   There are some gray areas. For example, if an automobile bursts into flames when accidently hit by a driver who could not stop at the traffic light, the automobile may be considered inherently dangerous. One can argue that the accident was caused by human error and the product did not contribute to the accident. A jury may not agree; it may decide that such accidents are normal events and the automobile should not go into flames in normal use.

2. *It minimizes the likelihood of danger:*   In other words, it minimizes the frequency of accidents. This can be achieved by designing products with higher reliability.

3. *It uses state-of-the-art technology for minimizing hazards as much as possible:*   In the example in item 1, a steel plate between the fuel tank and the driver should prevent an explosion.

4. *It minimizes life-cycle costs:*   Consider the cost of a steel plate at $11 per automobile. If millions of cars are produced the total costs will run into many millions of dollars. If the costs are prohibitive, one should evaluate alternatives such as adding a plate made of a tough plastic like Lexan or lining the gas tank with plastic. Usually a life-cycle cost analysis (Chap. 4) will reveal that a costly de-

sign change will produce lower life-cycle costs for at least one of the alternatives.

Caution: Safety decisions are not based on minimizing life-cycle costs alone. They should be based on all four goals above.

Example 5.2   What are the benefit-to-cost ratios for each design option below. The data apply to prevention of the exploding car accident above.

Option 1:   Install a steel plate. Cost $14. This will prevent all explosions.

Option 2:   Install a Lexan plastic plate. Cost $4. This will prevent 95 percent of explosions.

Option 3:   Install a plastic lining inside the fuel tank. Cost $2. This will prevent 85 percent of explosions.

The following data apply to the current design:

Possible fatalities from vehicles already shipped: 180

Expected cost per fatality: $500,000

Number of injuries expected (no fatality): 200

Cost per injury: $70,000

Expected number of vehicles damaged (no injury): 3000

Cost to repair the vehicle: $1200

Number of vehicles shipped: 6,000,000

solution   The cost for each option is the cost of implementing the change. The benefits are in terms of lives saved and avoidance of injury and damage.

Option 1:

$$\text{Cost} = \$14 \times 6{,}000{,}000 \text{ vehicles} = \$84{,}000{,}000$$

$$\text{Benefits} = (180 \text{ lives saved})(\$500{,}000) + (200 \text{ injuries prevented})$$
$$\times (\$70{,}000) + (3000 \text{ damaged vehicles prevented})(\$1200)$$
$$= \$107{,}600{,}000$$

$$\text{Benefit/cost ratio} = \frac{\$107{,}600{,}000}{\$84{,}000{,}000} = 1.2809$$

Option 2:

$$\text{Cost} = \$4 \times 6{,}000{,}000 = \$24{,}000{,}000$$

$$\text{Benefits} = (95\% \text{ accidents prevented})$$
$$\times [(180 \text{ fatalities})(\$500{,}000) + (200 \text{ injuries})$$
$$\times (\$70{,}000) + (3000 \text{ vehicles})(\$1200)]$$

Benefits = .95 × $107,600,000

$\qquad$ = $102,220,000

$$\text{Benefit/cost ratio} = \frac{\$102,220,000}{\$24,000,000} = 4.2592$$

Option 3:

$$\text{Cost} = \$2 \times 6,000,000 = \$12,000,000$$

Benefits = (85% accidents prevented)

$\qquad$ [(180 fatalities)($500,000) + (200 injuries)

$\qquad$ ($70,000) + (3000 vehicles damaged)($1200)]

$\qquad$ = .85 × $107,600,000 = $91,460,000

$$\text{Benefit/cost ratio} = \frac{\$91,460,000}{\$12,000,000} = 7.6217$$

Option 3 has the highest benefit/cost ratio. As noted earlier, however, the decision cannot be based on this figure of merit alone. From a moral standpoint, option 1 is the best. Actually, option 3 has some hidden costs which were not considered here, such as loss of future customers. But this option is better than not taking any action at all.

5.2.2 Safety engineering tasks

Several analyses can be applied as the design progresses. A typical PERT chart of tasks is shown in Fig. 5.2. It is important to recognize that as much as possible the tasks should be performed according to principles of concurrent engineering. This assures that system safety is consistent with reducing costs and enhancing schedules. Safety must always make business sense. Therefore, hazard analysis techniques form the core of system safety methodology. They allow designers to make engineering changes very early when they are least costly to implement.

5.2.3 Preliminary hazard analysis

Preliminary hazard analysis (PHA) is done at the concept stage so that safety considerations are included in tradeoff studies during early design. The purpose is to obtain initial risk assessment of a concept and to make sure the conceptual framework is sound. Time and money are wasted if the concept has to be changed later. Besides, it may be too late to change the concept. Many analysts find it difficult to conceptualize the hazards, but a systematic process should make it easier.

The depth of analysis always depends on the objectives and on the

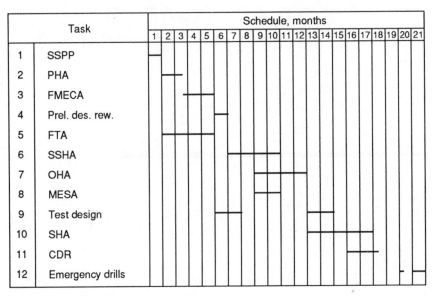

Task		Schedule, months																				
		1	2	3	4	5	6	7	8	9	10	11	12	13	14	15	16	17	18	19	20	21
1	SSPP																					
2	PHA																					
3	FMECA																					
4	Prel. des. rew.																					
5	FTA																					
6	SSHA																					
7	OHA																					
8	MESA																					
9	Test design																					
10	SHA																					
11	CDR																					
12	Emergency drills																					

Figure 5.2   A PERT chart for safety tasks.

severity of the accidents. The objective in PHA is to identify the hazards. As a minimum, the hazards that will impede the mission objectives must be identified. Such hazards are not limited to hardware and *at system level* include:

Hardware hazards

Software hazards

Procedural hazards

Human factors

Environmental hazards (including health hazards)

Interface hazards

The PHA process can be demonstrated with a chemical plant example (Fig. 5.3). The plant in this example stores a chemical in an underground storage tank from which the daily requirements are delivered to a unit tank. The chemical, if exposed to humans, is deadly. In addition, it can react within itself and cause an explosion; it also reacts violently if it comes in contact with moisture or water. The plant contains some safety features. Temperature and inventory sensors monitor the heat and the quantity of the chemical, respectively. A pressure monitor is included as a precaution. If the pressure is too high, a pressure-relief device vents small amounts of gas, which are

Figure 5.3   A chemical plant.

piped through a scrubber and neutralized. In addition, a flare unit can burn off most of the gas from a major leak.

PHA is basically a brainstorming technique, but some kind of organized approach helps in starting the process. Some system-level questions can be

Are there any hardware hazards?

Are there any software hazards?

Are there any human-induced hazards?

Are there any procedure-related hazards?

Are there any hazards to the public environment such as noise, radiation, or contamination?

Are there hazards from the existing environment?

Are there any obvious interface hazards among software, hardware, and humans?

Are there hazards due to materials such as lasers, propellants, and fuels?

The literature is not clear on how to identify hazards, but the author has practiced two more approaches very successfully. Use of both approaches in addition to application of the questions above provides a complete framework for PHA.

Approach A: Controlling the energy. According to method A, if certain energies in the system are controlled, there will be no accident. The following energies can be present in the chemical plant in Fig. 5.3:
Chemical energy:

Too much reaction

Chemical reacting with the tank material

Moisture in the tank

Pipe corrosion, tank corrosion

Gas leaks

Thermal energy:

High temperature in the tank

High temperature outside the tank

Pressure energy:

High pressure in the tank

High pressure in the pipes

Mechanical forces:

Vibration in the pipes

Vibration in the tank

Electric energy:

High voltage in the cables on control panels

Once these energies are identified, the following checklist can be applied. The checklist is a 10-point precedence test proposed by Dr. William Haddon.[2]

1. Prevent the marshaling of the energy in the first place.
2. Reduce the amount of energy marshaled.
3. Prevent release of energy.
4. Modify the rate or spatial distribution of energy.
5. Separate the energy release in space or time.

6. Do not separate in space or time, but interpose a material barrier.

7. Modify the contact surface or basic structure.

8. Strengthen the structure.

9. Move rapidly in detecting energy release and counter its extension.

10. If energy is beyond control, then stabilize the altered condition.

Approach B: Exploring the life-cycle phases. This method uses systematic prediction of possible hazards during the entire life cycle of a product. For each life-cycle phase, each hazard is assessed and controlled if possible:

Design: Are there hazards in the present design? Could there be any design mistake to consider?

Manufacturing: Are there any hazards if the product is made as intended? Could any manufacturing mistakes be made that will result in a hazard?

Testing: Are there any hazards during testing the product? Could any hazard be created during testing?

Storage: Are there any hazards in the way we plan to store the product? Could any one make mistakes in storage leading to unsafe conditions?

Handling and transportation: Are there any hazards if the product is handled and transported as intended? Could someone introduce hazards using unintended procedures?

Operations and use: Are there any hazards during intended use? Could someone introduce hazards by misuse or use in a different manner?

Repair and maintenance: Are there any hazards present during normal repair and maintenance? Could someone introduce hazards by mistake?

Emergency situations: Are there hazards in a situation such as emergency shutdown of an operating system? Can someone make mistakes because of unusual work pressure?

Disposal: Are there hazards in disposing of the product as intended? Could someone use different disposal procedures and introduce hazards?

An example of PHA format is provided in Fig. 5.4. At PHA level, it should be noted, major design improvements can be and should be made. These improvements are shown in the last column of all hazard

Preliminary hazard analysis

Hazard	Cause	Effect	Criticality	Recommendations	Final action
1) Temperature too high	Gauge malfunction	Gas leak	I	Provide cooling jacket on tank or redundant gauge	Provide cooling jacket
2) Too much inventory	Sensor malfunction	Explosion	I	Install redundant sensor	Install redundant sensor with automated control
3) Leaks through pipe joints	Vibrations, corrosion	Gas leak	I	Install stainless steel pipes	Install stainless steel pipes with flexible joints

Figure 5.4 PHA matrix for the chemical plant of Fig. 5.3.

analyses. In the PHA in the figure, the design changes are such that the whole design philosophy has to be changed. The existing design is a series structure (see Chap. 3). The improved design is a fault-tolerant structure involving redundancies.

A fault-tree analysis (explained later in this chapter) for the system-level failures can also help in identifying potential hazards. Other techniques are cause-effect diagrams (Chap. 6), design reviews, and simulation.

Hazard control criteria. Figure 5.5 shows the design criteria for controlling potential hazards. They apply to all hazard analyses. They are presented in order of precedence and can be summarized as:

1. *Design the hazard out* of the product. If the hazard cannot be eliminated, minimize the residual risk.

2. *Design for a fail-safe default mode* by incorporating safety devices or fault-tolerant features.

3. *Provide early warning* through measuring devices, software, or other means. The warning should be clear and should attract the attention of the responsible operator.

Figure 5.5   System safety design criteria.

4. *Implement special procedures and training* when the above means are unable to eliminate the hazard.

5.2.4 Subsystem hazard analysis

The literature is very scarce in the important area of subsystem hazard analysis (SSHA). It emphasizes the importance of SSHA, but rarely the "how" of it. SSHA is a detailed extension of PHA. For less complex equipment, where it is not necessary to break down the system into subsystems, it may simply be called *hazard analysis*. SSHA is based on the premise that an accident is a result of at least two events (a hazard and a trigger event) and can be prevented as long as either event is controlled. If neither of the two events can be prevented, then it is prudent to perform a fault-tree analysis to reduce the number of causes leading to the accident.

To perform subsystem hazard analysis the entire system is divided into subsystems. The subsystems themselves usually are complex enough to be considered systems. As an example, consider a mass transit subway system. The system mission is to transport the public safely underground. The subsystems may be defined as follows:

Trains

Station structures, including escalators and elevators

Tunnels

Central control room

Track and controls

Provisions for the elderly and handicapped

Communications

Fare collection

For each of these subsystems, a separate analysis should be performed. Each component or unit of the subsystem should be analyzed. In Fig. 5.6, the units of a fuel control subsystem of an automobile are shown. The subsystem hazard analysis for this subsystem is in Fig. 5.7. One should consult the bill of material to determine what to analyze. It is easy to miss important components such as adhesives, lubricants, and cables; high-voltage cables have been known to cause deformities in animals and farm products, while adhesives can be toxic. The analysis steps for each unit are

1. Identify hazards.

2. Identify the trigger event leading to an accident.

Figure 5.6 Units of a fuel control system.

3. Classify the criticality.

4. Identify the hazard or trigger control action.

The corrective action is determined by the criticality rating. For category I and II hazards, the mitigation is accomplished by improving the design. Inspection, testing, warnings, and procedural actions are not acceptable unless one is willing to accept the risk. If a design improvement is not obvious, it is necessary to perform fault-tree analysis (Sec. 5.2.5) or the equivalent to justify why one cannot change the design. Many times the FTA will lead to a reduction in frequency or may provide a complete solution.

5.2.5 Fault-tree analysis

Fault-tree analysis (FTA) is an inductive process especially useful for analyzing category I and II hazards, when the hazards have not been resolved. If the hazards can be resolved by design, the fault-tree analysis may not be necessary. The purpose is to identify the causes of hazards so that an effort can be made to eliminate as many causes as possible.

FTA can also be used for analyzing any problem: technical, management, or administrative. It can be used for manufacturing process control, human factors, and software mishaps. Some perform FTA at system level to identify many hazards.

FTA is a cause-and-effect diagram which uses standard symbols developed during the Minuteman missile program. It is also called root-cause analysis. A sample fault-tree analysis is shown in Fig. 5.8. The definitions of symbols commonly used are shown in Fig. 5.9.

Subsystem Fuel control
System 'x' vehicle

Subsystem Hazard Analysis

Part Name/ part number	Hazard description	Accident (trigger event)	Criticality	Recommended control	Revised criticality
Fuel Tank	Tank rupture	Differential in pressure between the inside and outside of the tank	I B	Qualify tank for 20-year ability to withstand rupture forces. Use ductile material.	III E
	Failure of tank seal	Any kind of flame or spark	I B	Provide double seal.	II C
Fuel pump	Fire	Spark in the area	II A	Mount fuel pump inside the tank.	III E
Fuel line	Rupture	Corrosion sufficient to penetrate through the wall thickness	II C	Use stainless steel lines or corrosion-resistant material.	II E
Filter	Dirty or clogged filter	High buildup of pressure	I C	Design for pressure relief. In addition, design filter for twice the current life.	III B
Carburetor	Fuel leakage	Vehicle rollover	II D	Use electronic fuel injection.	III D

Figure 5.7 Subsystem hazard analysis.

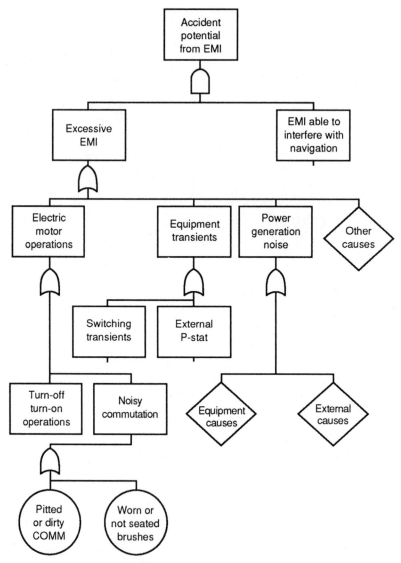

Figure 5.8 Fault tree for electromagnetic interference on an aircraft.

Construction of the tree in Fig. 5.8 started with a logical question: How can electromagnetic interference (EMI) cause an accident? Remembering that an accident is always a result of a hazard and a trigger event, an engineer constructed the second level of FTA. Here the hazard is "excessive EMI" and the trigger event is "EMI able to interfere with navigation." The AND gate symbolizes that *both* of these events must occur for an accident to occur. The branch "excessive

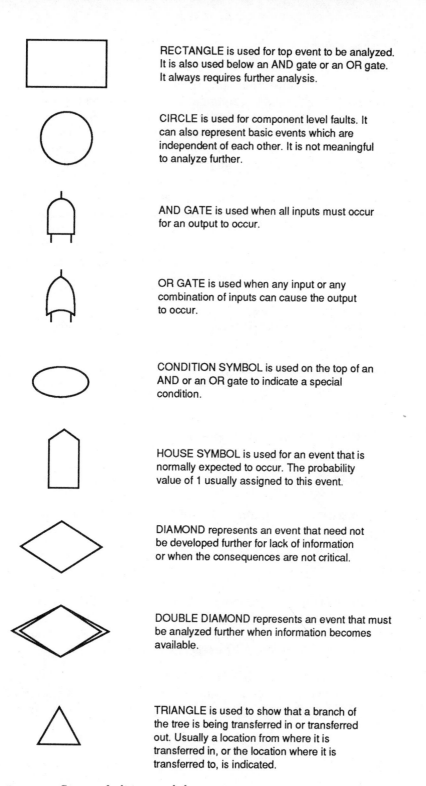

RECTANGLE is used for top event to be analyzed. It is also used below an AND gate or an OR gate. It always requires further analysis.

CIRCLE is used for component level faults. It can also represent basic events which are independent of each other. It is not meaningful to analyze further.

AND GATE is used when all inputs must occur for an output to occur.

OR GATE is used when any input or any combination of inputs can cause the output to occur.

CONDITION SYMBOL is used on the top of an AND or an OR gate to indicate a special condition.

HOUSE SYMBOL is used for an event that is normally expected to occur. The probability value of 1 usually assigned to this event.

DIAMOND represents an event that need not be developed further for lack of information or when the consequences are not critical.

DOUBLE DIAMOND represents an event that must be analyzed further when information becomes available.

TRIANGLE is used to show that a branch of the tree is being transferred in or transferred out. Usually a location from where it is transferred in, or the location where it is transferred to, is indicated.

Figure 5.9 Common fault-tree symbols.

EMI" shows four elements under an OR gate. Any of the four elements [electric motor operations, equipment transients, power generation noise, other causes (secondary faults)], or any combination of them, can produce excessive EMI. The process of analyzing causes for each event is carried out until further analysis is not meaningful. In a hardware tree the analysis may stop when failure causes are traced to component faults which are represented by circles. Similarly, circles may be used for the human or software details when the root causes have been postulated.

Usually a tree cannot be fitted in one page. A triangle is then used to denote that a portion of the tree is somewhere else. The location of this portion is indicated in the triangle. A triangle is also used for a branch which is repeated in several places.

Identifying causes in a fault tree. Conceptually one needs to have a systematic method of developing the tree. At least eight factors, shown below, can provide clues about the causes:

1. Can any equipment failures contribute to this effect?

2. Any material faults?

3. Human errors?

4. Could methods and procedures contribute?

5. Could software performance be a problem?

6. Could maintenance errors or the absence of maintenance contribute?

7. Could inaccuracies or malfunction of the measuring devices contribute?

8. Could environments such as chemicals, dust, vibration, shock, and temperature contribute?

Computing probability of an accident. If the fault tree contains the estimates of component probabilities, then the probability of accident can be computed by applying two fundamental rules to AND and OR gates starting at the bottom of the tree:

1. *Multiplication rule:* When an output results because of two or more input events occurring at the same time, as in an AND gate (Fig. 5.9), the output probability is computed by multiplying the input probabilities:

$$P(\text{Output}) = P(1) \times P(2) \times P(3) \cdots P(N)$$

This rule assumes the input probabilities are independent of each other. For example, carburetor and the radiator failures in an automobile are independent of each other because if one fails, the other is not affected. If a component fails as a result of the failure of another component, its probability of failure is not independent. Then a new equation has to be written on the basis the relationships. Reference 3 covers this situation. However, most fault-tree components are statistically independent of each other. This computation will rarely be required.

2. *Addition rule:* For an OR gate, where the occurrence of any input causes the output to occur, the probability calculation is based on whether the events are mutually exclusive or not. If two events are mutually exclusive, such as high pressure and low pressure (if one occurs, the others cannot occur at the same time), the probabilities are simply added up. Then

$$P(\text{Output}) = P(1) + P(2) + P(3) \cdots P(N)$$

If the events can occur at the same time (if they are not mutually exclusive, such as high temperature and high pressure) then

$$P(\text{Output}) = 1 - [1 - P(1)][1 - P(2)][1 - P(3)] \cdots [1 - P(N)]$$

If the probabilities are small (less than .001), they can be added together like probabilities of mutually exclusive events. The answer will be very close. Since probabilities in the fault trees are very low, most practitioners disregard the distinction between mutually exclusive and not mutually exclusive.

5.2.6 Cut set analysis

Fault trees can be put to many uses. One such use is identifying design weaknesses and strengths. A single event or component that can cause the system to fail is a potential weakness in the system. A tool called *cut set analysis* allows identification of all single-point failures and other paths which lead to a failure.

A cut set is a set of basic events whose occurrence causes the system to fail. For a hardware system it is a set of component failures leading to a system failure. A cut set is called minimum cut set if it cannot be reduced and can still cause the system to fail. Barlow and Proschan[4] developed an algorithm, which is available in several commercial computer programs, for reducing a cut set to the minimum.

This algorithm is illustrated in Fig. 5.10. Gates are labeled G0, G1, G2,..., GN. Numbers 1, 2, 3,..., n represent the other basic events. The algorithm begins with the top event, G0. The elements (G1, G2) of

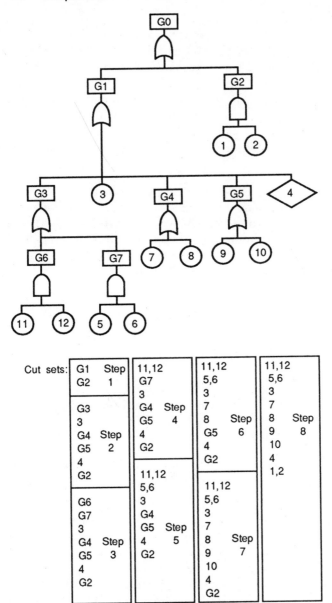

Figure 5.10 Cut set analysis.

this gate are connected by an OR gate, and are entered in separate rows in the matrix in the figure. (If the elements were connected by an AND gate, they would be entered in the same row.) All the gates are entered in the same manner until no more are left. This completes the analysis. The final analysis shows all the possible cut sets. A step-by-step walkthrough follows below.

The matrix in Fig. 5.10 shows eight steps. The first step shows gate G0 replaced by G1, G2 in *separate rows* because they are connected by an OR gate, as noted above. In step 2, gate G1 is replaced by its elements G3, 3, G4, G5, and 4. (Gate G2 is not replaced yet.) Step 3 shows G3 replaced by its elements G6, G7. (Other elements are untouched.) In step 4, gate G6 is replaced by its elements 11, 12 in the *same row* because they are connected by an AND gate. (The remaining elements are untouched.) The analysis continues in this way until step 8, when no gates are left. Each row is then a cut set. In this case, since the basic events are not repeated in the other sets, each row is a minimum cut set. Events (or components) 3, 7, 8, 9, 10, and 4 are single-point failures which can cause the system to fail. It is here a designer should consider redundancy or fail-safe features.

A similar tool called *minimum path set analysis* is an extension of this analysis. A *path set* is a set of components (or events) which, if it works, assures the system will work. A path set is a minimum set if it cannot be reduced further. The procedure is to change the OR gates to AND gates, and vice versa, in a fault tree and use the cut set algorithm. Each row in the final step is a path set. If any of these paths work, the system will work. An example of this analysis is in Ref. 4.

5.2.7 Failure-mode, effects, and criticality analysis

Failure-mode, effects, and criticality analysis (FMECA) evaluates the effect of component failures and is usually done by reliability engineers. (For details, see Chap. 3.) It would be better done by safety engineers, reliability engineers, and design engineers together, however. The reliability engineers may not pay detailed attention to safety-critical failure modes, and they often will not consider a failure critical when redundancies in design are present. A safety engineer may still consider such a failure a dangerous situation. For example, if an aircraft device has a redundant circuit board, a failure may not be critical if only one circuit board fails. But if the failed board is not detected and replaced immediately, another failure could be catastrophic. The safety engineer in this case has to assure that the repair action will be performed immediately.

Assessing the consequences of failure requires special training.

Many reliability engineers do not go to proper depths. For example, if a motor for an escalator on a train station fails, a reliability engineer may not relate the stopping of the escalator to a safety hazard, but a safety engineer knows that this can cause accidents. If an escalator comes to an abrupt stop, people fall on each other. A good design would require the escalator to come to a gliding stop. The least a safety engineer should do is let reliability engineers complete the FMECA and then review the effects of failures independently.

5.2.8 Maintenance engineering safety analysis

Maintenance engineering safety analysis (MESA) should first be done in the design phase, even though some may claim that maintenance procedures are not known until the system is in place. If one waits for the system to be installed, it is too late. If maintenance hazards are to be prevented, the system must be designed accordingly.

The difficulty of not having procedures is overcome in the following manner. A rough draft of the potential procedures is constructed with the help of experienced maintenance technicians who have worked on similar systems. Usually, at least 80 percent of the tasks are similar. This knowledge is supplemented by a method called technique of human error rate prediction (THERP).[5] This consists of analyzing maintenance safety on the basis of interviews with operating and maintenance personnel on similar systems. The analysis is updated as the design details become available. The best time to start this analysis is when preliminary design is approved. It must be complete before the critical design review. This is the last opportunity to make design changes.

MESA consists of writing the procedure in logical tasks and then treating each task as a component for hazard analysis. As an illustration, take three tasks in sequence for a proposed preventive maintenance procedure for a high-voltage transformer.

Task 1:   Shut off the power.

Task 2:   Check the oil level with a dipstick (a metal rod).

Task 3:   Remove an oil sample for chemical analysis.

In task 1, the following hazards are possible:

- The technician may think the power is off, but it is really on because of feedback loops.
- The power may be off, but there is a residual high voltage.
- Someone may turn on power at the source.

The task 2 hazards are

■ The technician can be electrocuted through the metal dipstick if the power is accidentally on.

■ Contact with transformer oil can cause toxic effects.

The analysis is shown in Fig. 5.11. It is interesting to note the design improvement for task 2. To prevent exposure to high voltage, use of a sight gage is suggested (instead of a dipstick) so that oil levels can be observed remotely. This action also prevents an accident from task 1, since it is no longer necessary to shut down the power. The oil sample can be removed through an insulated pipe without shutting down the power. Figure 5.12 shows a format suggested by the U.S. Navy.[3]

5.2.9 Event trees

Event trees are complements to fault trees. They trace the effect of a mishap and lead to all possible consequences. This information is required to develop training for emergency drills. In large systems, it is imperative that a mock accident be created periodically to see if people know what to do during emergencies and disasters. This is an excellent opportunity to check whether the evacuation plans are realistic. Such drills reveal many weaknesses in the system which can result in multiple fatalities. Figure 5.13 is an example of an event tree.

The procedure for constructing event trees is to visualize two sides, positive and negative, for each event. For example, the tree in Fig. 5.13 shows that an alarm may work or it may not. If it works, there is no damage done. If it does not, then there is a chance of manual detection; a person may detect the leak or may not. The analysis is carried on until all the consequences are considered. Event trees can also be used for estimating the probability of each consequence, since the sequence of events is known.

5.2.10 Operating and support hazard analysis

Operating and support hazard analysis (O&SHA) is done for operating procedures and support functions such as:

■ Production

■ Testing

■ Deployment

■ Handling

| Procedure title | Transformer preventive maintenance | Maintenance Engineering Safety Analysis | | | Date 3/16/89 Page 1 of 25 |
| System | Power distribution | | | | Analyst John Doe |

Task number	Task description	Hazard	Cause of hazard	Criticality	Recommended control	Revised criticality
001	Shut off power.	Power may not be cut-off.	Someone may accidentally turn on the switch, or the circuit may malfunction.	I C	Provide postive cut-off switch locally within the equipment.	Hazard no longer exists.
002	Check the oil level with a dipstick.	High voltage shock through metal dipsticks.	Same as in task 001.	I C	Provide a sight gage so the need for this task is eliminated (Note: task 001 is no longer required.)	Hazard no longer exists.
003	Remove an oil sample for chemical analysis.	Electrocution if power is on accidentally.	Oil conducts electricity.	I C	Provide a sight gage which may indicate contamination. Alternatively provide an insulated pipe through which oil can be removed without shutting off the power.	Hazard no longer exists.

Figure 5.11 Maintenance engineering safety analysis.

System Surface Missile
Subsystem _____

Maintenance Hazard Analysis (LIHA)

Equipment	Maintenance type	Maintenance function	Hazard	Hazard class	Safety feature or recommendation	Remarks/recommendation
1	2	3	4	5	6	7
CW transmitter	Corrective maintenance	Remove and replace high voltage power supply	Access to HVPS will activate safety interlocks causing HV to be discharged. Activation of battle short during this maintenance action bypasses the interlocks, creating a serious personnel hazard.	IV	Cabinet has an inadequate safety interlock design capable of being bypassed during battle short mode operation. Should have method positively assuring no voltage when desired.	Recommended power interrupt switch located on the cabinet to inhibit all power including battle short, during maintenance. A holding relay circuit could also be considered for this purpose.
Launcher	Preventive and corrective maintenance	All maintenance or work performed on or near the launcher.	Personnel working in or near the launcher are susceptible to injury from the launcher rails during remotely controlled train and elevation.	III IV	Locking pins are adequate when not in an alert status. A warning alarm is installed at the launcher to warn personnel of movement. Sound power telephone outlet for safety watch.	A launcher train circle should be provided to indicate danger zones. Ref. ship installation specification.

Figure 5.12 Maintenance hazard analysis from a U.S. Navy example.

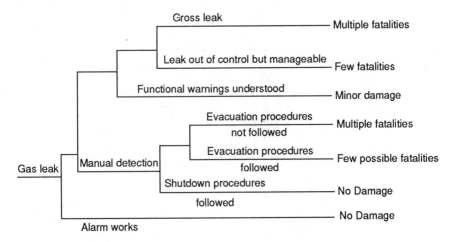

Figure 5.13   Event tree for a gas leak.

- Servicing
- Storage
- Modification
- Demilitarization
- Disposal
- Emergency action

The technique requires identification of activities by a procedure similar to MESA. Maintenance activities can also be covered in this analysis. Since maintenance activities usually are too many, MESA is done separately.

The inputs for this analysis come from previously mentioned analyses in which many operating hazards are identified. Other inputs which can help are prototype tests, mock installations, emergency procedures, and interviews with operating and maintenance personnel.

The analysis should include:[1]

- Activities which occur under hazardous conditions, their time periods, and the actions required to minimize their risk
- Changes needed in functional or design requirements for system hardware/software, facilities, tooling, or support/test equipment to eliminate hazards or reduce associated risks
- Requirements for safety devices and equipment, including personnel safety and life-support equipment.

Subsystem (1) _____

Operation Safety Matrix   Activity (2) _____

System safety engineering analysis

OSM Item No. (3)	Task Description (4)	Hazardous element (5)	Potential accident (6)	Effect (7)	Hazard class. (8)	Safety requirements (9)	Procedure reference (10)

Prepared by _____

Date _____ Issue ___ Rev. ___

Figure 5.14   Operating and support hazard analysis.

- Warnings, cautions, and special emergency procedures (e.g., egress, rescue, back-out)

- Requirements for handling, storage, transportation, maintenance, and disposal of hazardous materials

- Requirements for safety training and personnel certification

Analysis shows the weaknesses and critical tasks where a hazard is possible. Such tasks can be highlighted by darker print, and flow charts can be constructed with darker blocks depicting safety concerns and dotted lines for nonsafety activities. Similarly, dark arrows can be used to show that certain sequences must be adhered to. A format for reporting operating hazards is shown in Fig. 5.14.

5.2.11 Occupational health hazard assessment

Occupational health hazard assessment (OHHA) identifies health hazards so that engineering controls can be placed, rather than make short-term fixes or depend entirely on persons to protect themselves. Items to be considered are

- Toxic materials such as poisons, carcinogens, and respiratory irritants
- Physical environments such as noise, heat, and radiation
- Explosion hazards such as concentrations of fine metal particles, gaseous mixtures, and combustibles
- Adequacy of protective means such as goggles and protective clothing
- Facility environment such as ventilation and combustible materials
- Machine guards and protective devices

5.2.12 Sneak circuit analysis

Often a complex product is designed in portions by several design engineers. It is easy to overlook certain paths and conditions, especially with electronics and software. The sneak circuit analysis (SCA) technique was standardized by Boeing in 1967 as a result of its effort on the Minuteman missile.[4] It requires a lot of effort and should be applied selectively.

A sneak circuit is a latent path or condition in a system which inhibits a desired condition or initiates an unintended or unwanted action. This can happen in even a mechanical system. It is not a result of a component failure, nor it is a problem created by electromagnetic interference. It is mainly a result of lack of visibility of all the details. The methodology[6] consists of identifying and analyzing topological patterns from detailed manufacturing drawings, installation schematics, and engineering schematics. A network of trees is constructed with the help of a computer program. Five basic patterns, shown in Fig. 5.15, are vulnerable. Of these, the H pattern is most vulnerable; about 50 percent of sneak conditions are found in it. Design engineers should avoid the H pattern.

The analytical portion consists of asking questions related to each pattern. For example:

Can a switch be open when a load is required?

Can a switch be closed when a load is not required?

Can the clocks be unsynchronized when they are required to be synchronized?

Does the label indicate the true condition? Is anything else expected to happen which is not indicated by the label?

Can a current flow in the wrong direction?

Figure 5.15   Topological pattern for sneak circuit analysis.

Can a relay open prematurely under certain conditions?

Can power be cut off at a wrong time because of an energized condition in the interfacing circuit?

Can a circuit loose its ground because of interfacing circuits?

Clardy[6] points out that there are four broad categories of sneak conditions.

1. *Sneak path:*   Allows current or energy to flow along a suspected path or in an unintended direction.

2. *Sneak timing:*   Causes functions to be inhibited or to occur at an unexpected or undesired time.

3. *Sneak label:*   On a switch or control device, causes incorrect actions to be taken by operators.

4. *Sneak indication:*   Causes ambiguous or incorrect displays.

Since Boeing software is proprietary, some companies have contracted work through Boeing. Others have developed their own software. Some perform analysis informally. One company put the design engineers of various circuits together in a room and encouraged them to trace the effects of each circuit on interfacing circuits. Another very large company traced all the effects of failures by FMECA in a space program and determined whether a circuit was fault-tolerant. It discovered several sneak conditions.

Caution: A question such as this may come up: What will happen to the interfacing circuits if a relay is closed suddenly because of electrical overload? If the effect is not easy to trace, the best approach is to try it out during a scheduled shutdown or on a simulator. Do not ignore the question because no one has the answer.

5.2.13 System hazard analysis

System hazard analysis (SHA) is the analysis of interface effects and interface integration. Results of subsystem hazard analysis are evaluated to assess impact on other subsystems and on the total system. Interfaces are of several kinds: hardware to hardware, hardware to software, and software to software. Human interfaces also belong in this analysis. All must be evaluated. Interface integration usually involves merger of the supplier system with that of the customer. There is no standard method, and the format for reporting varies according to the needs of the system.

In the beginning of a project, PHA serves as a rough SHA. PHA is then replaced by complete and detailed analysis. Techniques similar to SSHA can be used for SHA. Inputs from FMECA and SCA are specially valuable because they affect the entire system. Fault-tree analysis can also be a very efficient tool for SHA.

5.2.14 Safety margins in design

While analysis helps discover potential hazards, the residual hazards must have adequate safety margins. Several techniques exist for establishing safety margins. The designer has to choose one or more on the basis of judgment.

1. *Overdesign approach:* Often an engineer will increase the thickness of a material to ensure an adequate margin. This is not a good practice. It is uneconomical. More important, if a stress concentration such as a notch or a sharp radius is present in a material, a load is multiplied 10 or more times at that location. This load is sufficient to start a crack. The overdesign can only postpone the accident not prevent it. To prevent an accident, the notch and the radius should be eliminated at these sites. In many steel structures, stress concentration at a sharp notch results in brittleness in a

thick section, even though the material has ductile properties. The brittle section fails much faster than the ductile section. A thinner ductile material allows a crack to propagate for years before failure.

2. *Statistical approach:*   This approach is more sound but requires destructive testing. A stress and a strength distribution is constructed to compute the probability of failure. A failure is assumed when a component's strength is lower than the stress applied to it. Literature provides such an analysis for situations where the stress and strength have normal distributions. Weibull analysis, covered in Chap. 3, can be used for situations for other distribution shapes. The worst-case safety margin for normally distributed strength is defined as

$$\text{Safety margin} = \frac{\text{maximum stress} - \text{mean strength}}{\text{standard deviation of strength}}$$

This ratio should be at least 4. In the aircraft industry, this ratio is expected to be at least 8.

3. *Rule-of-thumb approach:*   Many construction projects, such as dams, bridges, and highways, make use of thumb rules. The load applied is usually about one-third the yield strength. Allowance for impact load is made with an additional 30 percent increase in strength (or reduction in load). Allowances are also made for other applied loads. This approach has worked for traditional materials because of contractors' years of experience. However, it has some gross weaknesses similar to the overdesign approach. It is not safe against stress concentrations.

4. *Damage-tolerance approach:*   This approach is common in the aircraft industry but can easily be employed by automotive, oil refining, and chemical industries. A large aircraft can often have more than 15,000 square feet of structure with thousands of tiny flaws and cracks, but they take years to grow to a size at which they can cause failure, and the material gives sufficient warning of an impending failure. Typically, an engineer uses models for crack growth and chooses materials so that the crack is visible before a failure takes place. But the task of inspecting several thousand cracks is tedious and hidden cracks under rivets and in the assemblies are often overlooked. This analysis is usually supplemented by tests for 2 lifetimes. The approach is not always successful; a tiny flaw in a General Electric engine, which resulted in a major United Airlines accident in 1989, took 18 years to result in a failure. The flaw was inside a casting and was not easily noticeable during inspections.

5.3   System Safety in Manufacturing

Manufacturing processes are vulnerable. They can introduce hazards into products and can be a hazard to working personnel. Analysis for occupational hazards created by processes is covered in Sec. 5.3.2. Section 5.3.1 covers hazards introduced in a product by manufacturing.

5.3.1   Determining safety-critical items

To assess the possibility of hazards being introduced during manufacturing, the method of producing the product should be analyzed as in maintenance engineering safety analysis. The steps of the manufacturing process are reviewed to determine where safety-critical error can occur. Each hazard is classified by severity rating. All category I and II hazards are called *safety-critical items*. If a product is shipped with these items uncontrolled, costly recalls can be expected. In one case, a component used in a smoke detector required chemical cleaning during manufacturing, and the cleaning solution had to be changed every 8 hours. A production operator erroneously thought that the solution was clean enough, and did not change it. As a result, components rusted in the field. The rust caused some alarms not to function. Under pressure from the Consumer Product Safety Commission, the company recalled all its smoke detectors. The company is no longer in business.

Safety-critical items are not limited to manufacturing processes. They can be a result of handling, packaging, and shipping. Vibration and shock during transportation can cause a product to malfunction. These processes should also be reviewed. Some organizations conduct process FMECA. This is one of the best sources of hazard identification.

5.3.2   Manufacturing controls for safety

Control of safety-critical items is generally the responsibility of the process engineer or the production foreman. But system safety engineers must make them aware of their responsibility. Some organizations label such items *critical* or *major* on drawings. Others highlight them in the process sheets.

The following general procedure is a guide for controlling hazards. It is illustrated in Fig. 5.16 and the steps below for a welded joint.

Step 1:   Identify the safety-critical item (strength of welded joint).

Step 2:   Identify the inputs to the process which, if not right, will result in a weak joint. (Time allocated for welding; chemistry of weld; material, especially the carbon content, which influences tem-

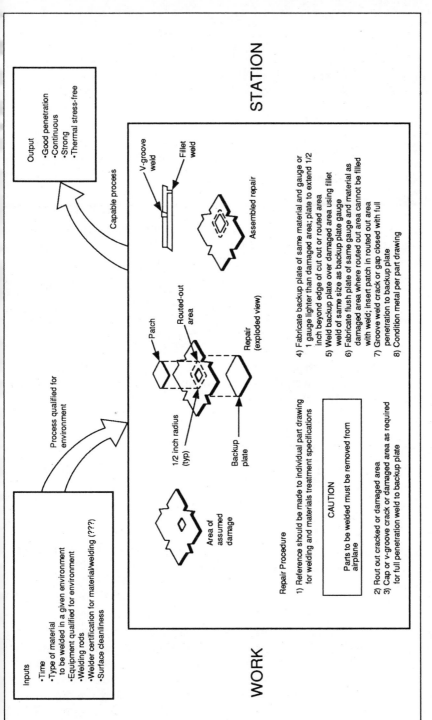

Figure 5.16 Manufacturing control of weld strength.

per qualities; equipment qualified for environment; quality of welding rods; welder certification; surface cleanliness; and storage environment of electrodes. *Note*: If electrodes are not dry, they will absorb oxygen, resulting in hydrogen bursts.)

Step 3:   Identify characteristics to control in the process itself (see *Caution* in the figure; other important characteristics such as cooling rate can also be identified.

Step 4:   Verify the output characteristics. If the output is not as expected, then either the inputs or the process need reevaluation. (The outputs for a welded joint are good weld penetration, continuous weld, strong weld, freedom from residual stresses, and ductility measured as Brinell or Rockwell hardness number.)

An example of what can go wrong in welding is the new aircraft hangar at a Chicago airport that collapsed when a commercial airliner flew over it. The failure analysis revealed that the welding was done on a very cold day. The faster cooling rate resulted in very brittle joints.

Generally, once quality assurance engineers are aware of the safety-critical items, they can help manufacturing in controlling them. See Chap. 6 for more techniques.

5.4   System Safety in the Test Stage

Most of this section is based on Ref. 7. The important principles and applications are covered here.

5.4.1   Testing principles

Accidents may result from unsafe conditions and unsafe events. In order to design effective safety tests, the analyst must know the potential unsafe conditions and events associated with foreseeable product uses. The purpose of testing is to assure that the product is capable of performing under these foreseeable situations without causing property damage or personal injury. The primary analytical tools used to assess potential accidents are the preliminary hazard analysis, failure-mode, effects, and criticality analysis, and fault-tree analysis.

In addition to the tools listed above, products must be analyzed for their ability to withstand the environments in which they are going to be used so that no unforeseen product or health hazard is created. To evaluate foreseeable environments of use, consideration must be given to conditions such as humidity, temperature, chemical vapors, dust, radiation, vibration, and shock. Safety tests are used to simulate the effects of long-term exposure to such environments. For example, products such as automotive seat belts degrade after such exposure and lose their ability to restrain passengers safely.

Some environments can create long-term health hazards. Other environments, such as a lack of oxygen in canned foods, are known to cause health hazards such as botulism. A recent controversy has arisen over blindness due to infection caused by lack of oxygen between the cornea and extended-wear contact lenses. Where issues of toxicity to humans exist, tests on laboratory animals are used. Drug companies routinely conduct laboratory experiments on test animals to determine the safety and efficacy of drugs before use by humans.

In addition to the analytical tools of system safety, special techniques have been developed by human factors specialists to evaluate the likelihood of human behavior and the effect of human error on safety. One technique is THERP (Sec. 5.2.8). FMECA can also be applied to each use and abuse situation to determine the impact of product component malfunctioning on product safety. Human factors analysis requires analysis of foreseeable misapplication and misuse. Tests can be developed to determine what human behavior will be under these situations and whether safety hazards will result because of human actions. Warning label verification testing, such as the program conducted by the National Electrical Manufacturers Association to develop the Mr. Ouch label to warn children of the hazard of electrical shock from electrical equipment, is an example of this type of testing. For some products, additional analytical tools may be developed or special tests required.

Assuming that hazards are not preventable by design, the criteria for effective safety tests is that the tests should demonstrate that, if the product can fail, it will fail in a safe manner and will not expose users or bystanders to dangers during its lifetime. Life tests can be developed to demonstrate safety over the life of the product.

5.4.2 Prerequisites for developing appropriate tests

Tests may be required to evaluate the safety impact of human actions, use, misuse, abuse, and repair and maintenance activities. The selection of appropriate tests depends on the potential hazards created during these actions. Therefore, thorough safety analyses are a prerequisite to effective safety testing. The safety analyses highlight areas in the product and human/product interactions that require testing. Depending on the nature of the product, preliminary hazard analysis;

failure-mode, effects, and criticality analysis; and fault-tree analysis may be performed.

Many times, PHA reveals the relationship between the design and the manufacturing process. A manufacturer of consumer products discovered that certain safety tests could be avoided if the tooling in manufacturing was changed. The improvements in tooling prevented production of out-of-tolerance parts that could have led to hazards. Such design and process improvements minimize the necessity for future safety testing. Because it may not be possible to eliminate some generic hazards by design changes, safety tests must be developed to evaluate the residual hazardous characteristics of the product. These tests will determine the likelihood of product failure under foreseeable use conditions, and the possibility of failure resulting in a hazardous condition. The tests also measure the severity of the hazard. Safety tests are also used to determine the effectiveness of safety devices, warnings, and interlocks in modifying user behavior and minimizing the likelihood of an accident resulting in injury or property damage. A study of water sprinklers showed that 3 percent to 5 percent of sprinklers, depending on the design, are not effective in actual fires.[8]

Before safety tests are developed, the causes of hazards should be minimized. Fault-tree analysis is one system safety technique that is used to systematically study the possible causes of failure. Each cause of failure is analyzed to determine if it is possible to eliminate the cause or change the product's design to decrease the likelihood of a failure. Tests to determine the response of the product to the various causes of failure are developed. Improving the design reduces the number of tests required to verify the safety of the product. The FMECA identifies many hazards not identified in the preliminary hazard analysis or the fault-tree analysis. Effective design of product safety tests depends on a knowledge of those components whose failures are most likely to result in hazards.

A procedure similar to preliminary hazard analysis can be used to evaluate the safety of maintenance and repair operations. At a minimum, human/equipment interface hazards must be analyzed. The working procedure rather than the product is tested in this instance. The designated procedures should create no hazards during product use or foreseeable abuse and misuse.

Design review is another tool used to evaluate the safety of a product. During the design review, the results of the safety analysis are discussed. Often, additional hazards are found to exist during these design review discussions. In order to practice reasonable care in design, the designer must try to eliminate these hazards. Durability tests should be designed for unpreventable hazards. A technique which can be used to evaluate the effects of faults is the event tree. Cost is an important consideration in determining the type and number of tests to be performed. A complex product may require many tests. The decision is based on the severity of the danger, its frequency, the cost of the improvement, and the availability of economically and technologically feasible alternative designs.

5.4.3 Product qualification tests

After the final design of a product is accepted by the design review team, products are usually tested to determine the type and frequency of failure over time, before production and sale to the consumer. These tests are known as *qualification tests*; they are designed to qualify the product for production. If the product exhibits unacceptably high failure rates, the design or production process is evaluated to determine what changes are necessary. Qualification testing is also performed when major changes are made to existing products, to determine how the changes affect the safe useful life of the product. A change meant to correct one product deficiency may result in other safety-critical failures.

Flaws appear in three stages of any product life cycle. Together they are known as the bathtub failure distribution. Every engineer should be familiar with the details of the distribution (see Chap. 3).

To simulate service life in the laboratory under controlled conditions, accelerated tests are required. Without higher stresses, the time required to test a new product might be many years. With accelerated testing, years of service life can be simulated economically in a much shorter time. Accelerated testing must be designed and performed with care. The engineering and physical properties of the test specimens must not change except at the point of failure. Stresses may range from 150 percent to 400 percent over those anticipated in actual service of the product. There are some precautions to be followed when performing accelerated testing. The failure modes at normal stress levels and those at accelerated stress levels should be the same. The approximate relationship between test time and field life of the product can be analytically predicted by the design engineer. Chapter 3 contains more details. These tests must be performed periodically to assure that the changes in manufacturing are not detrimental.

5.4.4 Production tests

Production tests are designed to detect variations in the manufacturing process. These variations include those caused by human error.

Deviations in the manufacturing processes can introduce safety-critical flaws in the product. The product may be improperly heat-treated. A brittle material is more likely to break in service, possibly injuring the user. A worker may forget to put a lockwasher or gasket in place. If these components are critical to the safe use of the product, production tests should be designed to detect their absence. Statistical design of experiments is useful in resolving many common problems. Special tests may be required for unique situations.

Production tests should, at a minimum, assure that the product conforms to engineering specifications, regulations, and standards. Absence of these tests may constitute negligence. The proper design of these tests is important to assure that they are able to find defects. Other production tests consist of testing for proper handling, packaging, and shipping. Tests should be conducted to evaluate the product's capability to withstand shock, temperature, and vibration during these activities. As with other tests, a thorough understanding of the potential hazards that can result from improper handling, packaging, and shipping is required in order to develop appropriate tests.

5.4.5 Tests for human-related failures

Human actions can introduce safety hazards. These can arise during intended use, abuse, or misuse. Tests should be designed to determine how the product will react under these foreseeable situations. It is often difficult to predict how the product will behave without sufficient testing under these conditions. One example involves the location of the horn in automobiles. Most automobiles used to have the honking mechanism located in the center of the steering wheel. Drivers become accustomed to pressing the center of the steering wheel in an emergency. However, some late-model automobiles locate the mechanism on the periphery of the steering wheel or on a separate lever behind the steering wheel. A person renting different automobiles in different cities, and thus unfamiliar with the location of the horn mechanism, may be unable to react appropriately in an emergency. Tests are performed to determine how suitable the nonstandard feature is under foreseeable conditions of use. Mockups of automobile interiors can be evaluated by human factors specialists through the observation of test drivers.

Tests are often required to determine foreseeable misuse and abuse. One toy manufacturer invites children to a company facility to use its products and observes how the children misuse or abuse the products, through one-way mirrors. A controlled test may not be adequate. Therefore, the company also provides toys for children to use at home. Parents are asked to record how the child uses the toy and to note possible safety problems. On the basis of these tests, the manufacturer can decide to alter the design of the product or to recommend parental supervision during product use. The manufacturer may also determine that only children above a certain age should be allowed to use the product. Without adequate testing, a jury may find that injuries caused by foreseeable use of the unmodified product are attributable to product defects.

For large, complex systems such as aircraft or chemical plants, companies use simulators to evaluate the effectiveness of the arrangement of controls and instrumentation and to train new operators. One method of testing deliberately introduces defects into a product to determine how the user reacts. Where testing is too dangerous for actual users, carefully designed dummies that simulate human responses can be used. Automotive crash testing of new vehicles makes extensive use of this method. A number of computer models have been created to simulate the motion of occupants involved in crashes to determine the crashworthiness of automotive designs.

5.4.6 Testing to prevent accidents during repair and maintenance

Tests to prevent accidents during repair and maintenance are similar to those used to determine human-related failures. The focus is on those persons who will be performing repair and maintenance. For complex products, these tests are expensive. Failure-mode, effects, and criticality analysis is one of the techniques that can be used to determine what can go wrong during repair and maintenance. Testing to verify the analysis should be conducted. Before the tests, a list should be created of the possible things that can go wrong. The test will usually reveal additional problems that were not predicted by analysis. For example, a manufacturer's analysis indicated a possibility of poor connections in underground cable assemblies. Because the cables were in the ground, a portable x-ray machine was used for checking poor connections. This nondestructive test revealed several other problems not anticipated during analysis. Once these additional problems are discovered, their risks can be determined and the acceptability of the design evaluated.

5.4.7 Testing the safety of design modifications

Often the solution of one safety problem creates another. Design modification made to reduce costs without testing for the effect of the modification can affect the safety of the product. Usually it is not necessary to test the complete product. Safety analysis is necessary to determine the extent of the required testing. The analysis can be in the form of a mini-design review to critique the effects of the change on durability, assembly clearances, chemical properties, physical properties, and failures during shelf life. This example is not unusual: A roof disintegrated into an almost muddy material because a supplier had increased the sulfur content of the roofing material. The sulfur reacted with the moisture in the air to cause an undesirable change in the finished product, the roof.

Changes in the production process should also be evaluated. The production process may affect the safe performance of the product.

5.4.8 Testing procedures

A product may not necessarily be hard goods. A repair manual containing procedures is also a product. These procedures must be tested for safety the same way that products are tested. As a product is made up of components, a procedure is made up of steps. Some steps can result in accidents; a wrench falling from a scaffold or ladder can injure a person below. Methods for testing procedures include job sampling, emergency drills, change analysis, and THERP.

Job sampling is generally applied to inspection processes. It can also be applied to the evaluation of procedures. Several users are observed to measure adherence to correct safety procedures. Observations may reveal unsuspected hazards in the procedures and may result in improvements to the procedures.

Emergency drills are used to evaluate complex procedures. For example, to determine whether passengers can evacuate an aircraft in 90 seconds, an emergency is simulated and the ability of the passengers to react properly is evaluated. This is a powerful tool because it evaluates whether people can respond to unforeseen hazards safely.

5.4.9 Analyzing test data—the right way

Engineers who do not have formal training in statistics are likely to make mistakes when evaluating test data. For example, the following test data were collected relating to the strength of a material (in pounds): 110, 120, 150, 150, 150, 160, 170, 170, 180, 190, 200, 400. The minimum strength required for this product is 100 pounds. Should the analyst accept this product on the basis of the test data? Each of the 12 test samples had a strength higher than the specification. If the data are plotted according to the Weibull distribution, it is discovered that there is 50 percent chance that approximately 4 percent of the units can fail below 100 pounds. Knowing these facts may alter the decision to accept the material. Because a sample size of 12 units was not large enough, the defective units did not show up in the test. When a larger sample of this product was tested, some defective units did show up. The lesson is that a manufacturer should not use data casually when dealing with safety. The probability distribution must be adequately analyzed before a decision is made. Chapter 2 contains details of this methodology.

5.4.10   How much testing is enough?

A manufacturer should test 100 percent of its products for safety if possible. Some companies inspect a product twice during the production process using two different inspectors. If the test is destructive, 100 percent testing cannot be performed, of course. In such cases, a statistician can determine the necessary sample size.

5.5   System Safety in the Use Stage

Once a system is installed, more hazards are discovered. They will require engineering changes. These changes must be reviewed for safety. Several tasks outlined below must be performed on an ongoing basis.

5.5.1   Closed-loop hazard management

The effort to resolve hazards permanently must go on. Hazards should be reported even if they did not result in an accident. For example, a nurse who trips on a wire in a hospital may not get injured. But later someone else may trip on the same wire and sustain an injury. Such occurrences are usually reported on an *incident report*. A *closed-loop system* is one that ensures that the wire will never again be in someone's path anywhere in the hospital. Someone must verify this in future audits.

The safety audit is a powerful tool to assure that hazards are removed. If possible, the audit should be conducted by an independent person from outside the audited area. The auditor should also check for hazards reported in previous audits.

5.5.2   Integrity of the procedures

In no case should anyone be allowed to violate safety procedures. A separate audit should be conducted to check their validity and that they are adhered to. Job sampling (Chap. 8) is a good technique for procedure audits.

5.5.3   Control of changes

There are two basic types of changes: (1) changes in the product design and (2) changes in the manufacturing process. These two changes cause many other changes in test procedures, inspection procedures, training material, and disposal procedures, for example.

Changes in product design.   To control changes in product design, configuration control procedures must be followed. *Configuration control* means preserving the functional integrity of the product; it does not mean merely controlling the documents. Every change must be reviewed to see that these products configurations are not affected:

- *Functional configuration*, which requires that the functional baseline will be maintained after a change.

- *Allocated configuration*, by which certain safety and reliability requirements are allocated to certain portions of the system to be compatible with another system. Integrity from system to system must be assured after a change.

- *Product configuration*, by which a product or a subassembly is required to have certain features to assure its form or its fit in the overall system.

Three major steps are carried out in a configuration control system. First, decide which of the three configuration controls above is appropriate (*configuration identification*). Second, assure that the changes are evaluated and implemented in all the affected documents (*configuration control*). Third, record and report the status of all the changes (*configuration accounting*).

Changes in manufacturing processes.   Changes in methods of production should be carefully evaluated. One efficient procedure[9] is *change analysis*. In this method the new procedure is written and compared side by side with the old procedure. The differences are identified. They are evaluated for their effects. Other techniques that can be used are fault-tree analysis and event trees.

5.5.4   Accident investigation

If a mishap has already occurred, failure analysis laboratories are utilized for establishing the failure mode. The tools described in Sec.

5.5.3 can also be used for accident analysis, especially change analysis procedure. For example, a mushroom packaging factory went from a manual assembly line to an automated line. Surprisingly, automatically packaged mushrooms were found to have botulism after two consumers died. Change analysis revealed that the only difference between the old and new processes was that the mushrooms were cut into smaller, flatter pieces in the new process. Investigation showed that when the flat surfaces of two mushroom pieces were joined, oxygen was driven out between them, and this caused botulism.

5.6 Topics for Student Projects and Theses

1 There are several hazard analysis techniques (FMECA, FTA, O&SHA, etc.). They seem to be done independently, and some are not done at all. Users do not seem to integrate them. Develop a better approach so that the hazard analysis is complete, useful, and done in the shortest possible time.

2 Hazard analysis is generally done during design phase; very rarely do operating personnel make use of it. Discuss how hazard analysis should be used for improving operational safety.

3 Literature is almost completely void as to how one can apply system safety principles in manufacturing. Conduct a literature search and develop recommendations.

4 Perform a literature search to summarize techniques for assessing safety margins. Briefly present your own recommendations.

5 Conduct a literature search to define durability. What definition would you recommend? Justify your views in detail.

6 Propose your own methodology for performing software hazard analysis.

7 Conduct a literature search to define software system safety. Propose and justify your own recommendation.

8 Develop a safety test plan for any product. Justify each test.

9 Perform O&SHA for a product. Recommend design changes for reducing the risks.

10 Develop a methodology that you feel is appropriate for risk analysis.

11 Discuss why safety in design is a good business practice. Attempt to justify your statements with quantitative examples.

12 Propose a method for considering human factors during product design.

References

1. *System Safety Program Requirements*, Mil-Std-882, 1984.
2. D. Raheja, "A Different Approach to Design Review," *Reliability Review*, American Society for Quality Assurance, June 1982.
3. *System Safety Engineering Guidelines*, NAVORD-OD 44942, Naval Ordnance Systems Command,
4. Richard E. Barlow and Frank Proschan, *Statistical Theory of Reliability and Life Testing*, "To Begin With," (Gordon Pledger, 1142 Hornell Dr., Silver Spring, Md. 20904), 1981.
5. W. Hammer, *Product Safety Management and Engineering*, Prentice-Hall, Englewood Cliffs, N.J., 1980
6. Robert C. Clardy, "Sneak Analysis: An Integrated Approach," *International System Safety Conference Proceedings*, 1977.
7. D. Raheja, "Testing for Safety," in Shepard's/McGraw Hill, *Products Liability: Design and Manufacturing Defects*, Lewis Bass (ed.), McGraw-Hill, 1986
8. "Automatic Sprinkler Performance Tables," *Fire Journal*, National Fire Protection Association, 60 Batterymarch Street, Boston, Massachusetts 02110, July 1970.
9. William Johnson, *MORT Safety Assurance Systems*, Marcel Dekker, New York, 1980.

Further Reading

Malasky, Sol W., *System Safety: Technology and Applications*, Garland STPM Press, New York, 1982.

Peters, George A., and Barbara J. Peters, *Automotive Engineering and Litigation*, Garland Press, 1984.

Roland, H. E., and Brian M. Moriarty, *System Safety Engineering and Management*, John Wiley and Sons, New York, 1985.

Weinstein, Alvin S., et al., *Product Liability and the Reasonably Safe Product*, Wiley-Interscience, New York, 1978.

Yellman, T. W.: "Event-Sequence Analysis", *Proceedings of the Annual Reliability and Maintainability Symposium*, 1975.

6

Quality Assurance Engineering

6.1 Quality Assurance Principles

Most life-cycle costs are locked in by the design philosophy, and quality is no exception. Like reliability, it is built into a product. On the other hand, this does not mean that process control in manufacturing is unimportant. Process control is of the utmost importance to prevent premature failures, which can slash an inherently high reliability to a dismally low value.

Quality is conformance to a set of requirements that, if met, results in a product that is fit for its intended use.[1] It is a broad term applied to customer satisfaction. It includes quality of design, quality of processes, quality of services such as on-time product delivery, and the value perceived by the customer. (The "customer" can also be the production operators and managers in the downstream departments.) Since the goal is customer satisfaction (internal as well as external customers) the overall focus should be on customer needs.

Traditionally, companies have aimed at reducing scrap and rework costs rather than satisfying customer needs. Apropos of this philosophy, Dr. W. Edward Deming once said, "Working hard will not help you if you don't do the right things." This principle emphasizes that the requirements will result in fitness for intended use only when engineering specifications truly reflect the expectations of the customer. If not, blind adherence to specification will be counterproductive. If customers are offered high-quality products at reasonable prices (the "right things"), then the reduction of scrap and rework costs will automatically be considered.

A good quality assurance system can be represented by the model in Fig. 6.1. The model applies to new products as well as existing ones; in both the cases, knowing the customer's needs is the key to success. For new products, the inputs can come from potential customers and from

Figure 6.1   Quality assurance model, with suggested ownership.

the design reviews. When the product already exists, the inputs come from field complaints and customer contacts.

The model emphasizes the following steps:

1. Use various sources to get customer inputs, including the reasons for lower sales.

2. Establish "ownership" (total accountability) for all customer concerns. Specific people must be made clearly responsible and accountable for them. There should be no "rock throwing" such as manufacturing blaming design and vice versa. Top management must be the leader, not the follower. For efficient quality assurance, ownerships must be coordinated so that product and process design and the maintenance development proceed simultaneously and coherently. This definitely requires not only commitment but also involvement by top management.

3. Make an effort to change the product design so that it is immune to variability and uncontrolled conditions in manufacturing. Alterna-

tively, make an effort to change the process to eliminate these problems permanently.

4. If a quality problem cannot be removed by step 3, implement process control measures. Similar measures should be worked out with the suppliers.

5. The process should be inherently reliable. Therefore, assess operating procedures and maintenance procedures for their ability to keep the process in a controlled state all the time.

6. Make quality audits to assure that the product meets its ppm (nonconforming parts per million) goal and that the customer is proud of it. The pride shown by the customers for your products is the best measure of quality!

7. Apply these efforts to product design, process design, and operating procedures concurrently and cohesively.

6.2 Quality Assurance in the Design Phase

Quality starts during specification writing with an informal review. Detailed analysis begins as soon as the first draft of the specification is issued. Chapter 10 describes the tasks and important milestones for quality assurance engineering. These are an integral part of the overall system assurance plan.

Two analyses are required at this stage. The first is often referred to as the *product design review*. Quality function deployment, benchmarking, applying the quality loss function, and use of checklists are some tools for the product design review. The second analysis is called the *process design review*, to assure that the anticipated process concept is sound. These reviews are part of the system assurance design reviews in which reliability, maintainability, safety, and logistics support are also determined.

The process must deliver a high yield, which is often defined in terms of parts per million levels. A process producing nonconforming parts at a level less than 100 ppm is looked on as a standard of excellence. Those who achieve this level of performance are those who control quality characteristics closest to the target value.

In some industries, of course, the state of the art is not fully developed and quality is difficult to predict. Nevertheless, during the design phase, even though one may not be able to accurately predict quality levels, one should be able to judge that some processes are more reliable than others. For example, an automated manufacturing process might produce more consistent dimensions than those produced manually by different machine operators.

It is never too early to think about process reliability. Unfortunately, it is this area which industry has failed to recognize as *the* major competitive tool. A process design is done once but the labor and material in manufacturing are wasted hundreds of times by a poor process design.

6.2.1 Product design review for quality

There are several design reviews during product development. They are updated in each stage as the details become available. Their results are used in development of operating procedures, repair and maintenance procedures, and training. As a minimum, the reviews are done at least at these milestones:

1. Preliminary design review
2. Detail design reviews (several)
3. Critical design review—to approve the final design
4. Preproduction design review—after the prototype tests
5. Postproduction design review
6. Operations and support design review

The involvement of quality assurance starts at the preliminary design review and becomes much larger as design becomes firm. The largest effort should go into the preliminary and the detail design reviews because about 95 percent of life-cycle costs are established at this time. Lack of a strong effort will increase these costs dramatically. The money saved by preventing losses is at least 10 times the money invested in these design reviews.

The function of quality assurance at the preliminary design review is to make sure the new designs do not have the quality problems of similar designs in the market. Quality assurance engineers therefore must know the strengths and weaknesses of the competitive products. The review should examine the product's ability to perform its function in all the foreseeable situations. Any problems or concerns that show up might require a change in the concept of the design. For example, the author participated in design of a robot that was supposed to pick up and cut thin metal lead frames for a very large scale integrated (VLSI) circuit manufacturing process. There was a good possibility the robot would not be able to perform its function if its timing was not synchronized with another robot which was to pick up paper packing between two lead frames. The design review team therefore decided to operate both robots from the same clock sequentially. This action changed the design concept.

Quality function deployment. A value analysis tool applied to product and process development, quality function deployment (QFD) can be used for developing test strategies and for converting requirements to specifications. The concept originated in Japan. For new product development, it is a matrix of customer needs versus design requirements. The inputs can come from many sources including interviews, market analysis, market surveys, and brainstorming. Vannoy and Davis[2] suggest using fault trees to identify products with negative characteristics and convert them to success trees to identify customer needs (see Chap. 5 for fault-tree methodology). For example, they identify the following needs of the customers for an automobile:

- Price in relation to expectations
- Expectations at delivery
  Performance
  Features
  Attractability
  Workmanship
  Customer service
  Safety
  Serviceability
  Conformance to regulations
  Perceived quality
- Expectations over time
  Reliability
  Effective preventive maintenance
  Less preventive maintenance
  Number of failures below expectation
  Availability of repair parts
  Lower repair expenses
  Expenses covered by warranty
  Reasonable repair time
  Better customer service
  Time to failure within expectation
  Type of failure within expectation
  Durability
  Performance
  Safety
  Customer support

Quality function deployment means turning these needs into design engineering requirements. Similarly, these needs can be translated to manufacturing controls, test programs, and customer service requirements. Reference 3 contains more information.

Benchmarking. Benchmarking is a process of measuring one's own products and processes against those of leaders in the field and setting goals to gain a competitive advantage.

The procedure[1] starts with identifying items to benchmark during product planning and identifying key engineering features. It continues with these steps:

- Determine what product features to benchmark against. The targets may be companies, industries, or technologies.

- Determine existing strengths by collecting and analyzing data from all major sources, including direct contacts.

- From each benchmark item identified, determine the best-in-class target.

- Evaluate your processes and technologies in terms of benchmarks and set improvement goals.

Actually, the best-in-class target is always a *moving target*. Benchmarking should take this into account and not just be satisfied with the current best-in-class. A leader is the one that arrives first in the market with new targets and establishes its product as the world standard. Examples are IBM mainframe computers, Sony pocket radios, and Intel integrated circuits.

Quality loss function. The concept of the quality loss function is based on the engineering principle that, if all components are produced very close to their target value, product performance can be expected to be the best, and the costs to society can be expected to be lower. Unfortunately, over the years manufacturing managers have dismissed the principle as too costly to implement, until Dr. Genichi Taguchi modeled it quantitatively. He convinced managers that the cost of quality goes up, not only when the product is outside specifications, but also when the product deviates from the target value within the specifications. The more the deviation, the higher is the expected life-cycle cost. Naturally, the performance over the long term will not be identical for parts produced at the lower and upper ends of the tolerance band. Their performance is not likely to be as good as that of products produced at the target value. Inadequate performance will result in higher warranty costs and higher in-house inspection costs. It will ultimately turn away customers.

To understand the quality loss function, it is necessary to understand Taguchi's philosophy, which Kackar[4] has summarized as follows:

1. An important dimension of the quality of a manufactured product is the total loss generated by that product to society.

2. In a competitive economy, continuous quality improvement and cost reduction are necessary for staying in business.

3. A continuous quality improvement program includes incessant reduction in the variation of product performance characteristics about their target values.

4. The customer's loss due to a product's performance variation is often approximately proportional to the square of the deviation of the performance characteristic from its target value.

5. The final quality and cost of a manufactured product are determined to a large extent by the engineering designs of the product and its manufacturing process.

6. A product's (or process's) performance variation can be reduced by exploiting the nonlinear effects of the product (or process) parameters on the performance characteristics. Details of this principle are contained in Sec. 6.2.3.

7. Statistically planned experiments can be used to identify the settings of product (and process) parameters that reduce performance variation.

The loss function concept is the application of a life-cycle cost model to quality engineering. Taguchi used a simple model of the loss imparted to the supplier, the customer, and society in general. It is oversimplified but serves its purpose in highlighting the fact that a product is cheaper and better if it is consistently produced close to its target value.

Figure 6.2 shows the model. T is the target value of a variable at which the product is expected to perform best. Values of the variable—which may be a dimension, for example—are plotted on the horizontal axis and the loss associated with each value of the variable is plotted on the vertical axis. The loss function is

$$L(y) = k(y - T)^2$$

where k is constant and $(y - T)$ is the deviation from the target value. This model assumes the loss at the target value is zero; that is, the customer is satisfied if a component is produced at the target value. It further assumes that customer dissatisfaction is proportional only to the deviation from the target. The value of k may be determined by

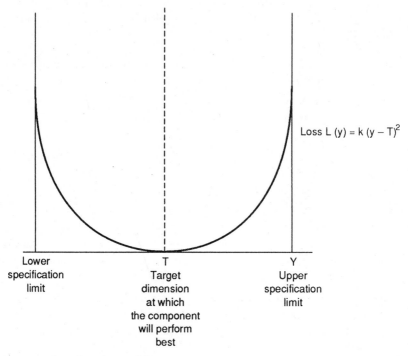

Loss L (y) = k (y − T)²

Lower	T	Y
specification	Target	Upper
limit	dimension	specification
	at which	limit
	the component	
	will perform	
	best	

Figure 6.2 Loss function model.

substituting an estimated loss value for an unacceptable deviation such as the tolerance limit. For example, if a Rockwell hardness number of 56 is the target value and if a customer's estimated loss for Rockwell hardness beyond 59 (the specification limit) is $200, k may be calculated as

$$k = \frac{\text{loss}}{(59 - 56)^2}$$

$$= \frac{200}{(3)^2}$$

$$= 22.22$$

The model may be written as

$$L(y) = 22.22 \, (y - T)^2$$

6.2.2 Process design review for quality and yield

During the preliminary design review, *all* the customer needs should be considered. As soon as the preliminary design is approved, the proposed flow chart for the process should be prepared. For some processes the analysis can be begun during the concept phase. It is advisable that the quality engineer work with the process engineer and the reliability engineer in a team. The purpose is to assure the process will work properly at the desired level of yield.

In designing for reliability, the team should quantify three factors and include them in the process specification. First, the process equipment must be capable of producing the right output (narrow distribution at the target value) consistently. Second, the process should be in operational condition as much as possible. Third, the process should be utilized to its full capability. These factors in the specifications can be respectively called the process effectiveness Y_p, process availability A_p, and process utilization U_p. Thus, process effectiveness can be defined as the probability that a process will work during the scheduled hours, that it will deliver the desired yield and that it will work at an intended speed, or

$$E_p = A_p \times Y_p \times U_p$$

Note: A standard definition of process reliability does not exist.

Process yield performance may be defined as the probability that a process will produce the part or assembly within the specified range of target values during the time until a scheduled maintenance function is performed. It is the ratio of good output to total output. If the specification calls for process yield performance equal to 99.99999 percent, it means the process deviation outside the target range will be less than 100 ppm. The equipment should demonstrate this capability in a statistical analysis of data over a reasonable length of time. *Process availability* can be defined as the ratio of total uptime to total scheduled time. If a process is used for two shifts a day for 5 days a week, then the total scheduled hours are 8 hours × 2 shifts × 5 days = 80 hours per week. If the equipment is in running condition for 64 hours, then the process availability for the week is $A_p = {}^{64}\!/_{80} = 0.80$. Process utilization is the ratio of actual production quantity during a given period to its designed quantity. It is sometimes called the *speed ratio*. If the industrial engineering standards require 200 assemblies to be produced in 8 hours, and if only 70 are actually produced, the process utilization is ${}^{70}\!/_{80} = 0.875$ or 87.5 percent.

The following examples demonstrate that quality effort must be concurrent with the design work.

Example 6.1 An integrated circuit manufacturer bought a wire bonding machine. When the parts from a typical production run were analyzed, it was found that about 0.5 percent (5000 ppm) of the bonds had low strength. A Weibull plot (Fig. 6.3) for 2370 bonds shows not only that 0.6 percent of the bonds were of poor quality but also that the machine produced two levels of bond quality, as

Figure 6.3 Analysis of wire bonding equipment.

indicated by the two slopes on the plot. (See Chap. 2 for Weibull distribution concepts.) Had the buyer conducted some experiments at the supplier's facility, the mistake of buying such equipment would not have been made. One may say that this kind of study might not be possible during the design phase because the proposed equipment is special. In such a case, demonstrations on similar equipment should be required at a potential supplier's facilities.

Example 6.2   A vacuum circuit breaker concept was approved but the design engineers demanded tight tolerances for the parts. The author, having had long experience with the suppliers, knew that the yield of such parts would be less than 60 percent. In other words, the suppliers would eventually increase the price at least 40 percent to make up for the scrap and rework. A decision was made to test some suppliers. It was shocking to learn that not only did suppliers raise the prices tremendously but also very few sent bids. The result was that the product idea was dropped before it became a reality. It is better to drop a product at this stage than to waste millions of dollars more and drop it later.

Capital equipment analysis.   In capital equipment analysis, an effort is made to minimize downtime by a proper critique of purchased equipment. One of the best known techniques for this is the equipment failure-mode, effects, and criticality analysis. This technique is essentially the same as design FMECA, but emphasis is on minimizing process downtime; the technique is not to be confused with process FMECA, which will be discussed later in this chapter.

Equipment FMECA uses standard design FMECA, covered in Chap. 4, and evaluates the effects on downtime. The main objective is to introduce design changes that will improve process availability. If the design fix is not effective, then reliability-centered maintenance (see Chap. 4) is introduced. As an illustration, consider the equipment FMECA format in Fig. 6.4. It considers the effect of downtime in terms of how long it will take to diagnose the fault and repair the device. It can be noted that the *action plan* column contains design or preventive maintenance action instead of inspection. Even preventive maintenance should be avoided if a design change is feasible. Inspection should be used only as a last resort.

Many will agree that performing equipment FMECA is a responsibility of the reliability engineer. The author, however, in over 25 years' experience, has rarely seen reliability engineers get involved in process equipment reliability. Since the process engineer is responsible for the process, the ownership should belong there. If the ownership is assigned to the reliability engineer, the process and the quality engineers should still be involved.

6.2.3   Design optimization for robustness

As design work progresses, quality assurance activities should take a foothold in the formal design reviews. Technical analysis should be done to assure that the product design is optimized and the process reliability, yield, and availability are maximized. Design engineers,

Failure mode	Failure cause	Effects				Criticality			Diagnosis time (h)	Repair time (h)	Expected lost time per month	Action plan
		Local	Sub-assm.	Machine	Other	Downtime	Freq.	S & F				

Figure 6.4   An equipment FMECA format.

manufacturing engineers, reliability engineers, and quality assurance engineers should jointly perform experiments, either on hardware or in software simulation, to assure interchangeability, insensitivity to noise, and the optimum tolerance for each component. These parameters and tolerance ranges should be scientifically established to optimize the performance so that it is robust in any operating and environmental conditions. Methods proposed by Shainin,[5] Taguchi,[6] and Hicks[7] should be used. Shainin's and Taguchi's are probably the most popular approaches to robust design.

The Shainin approach. Shainin uses a variable search pattern technique to isolate the important variables and interactions that govern product performance. The search pattern uses multi-vari charts and/or full factorial (all possible combinations) experiments to zero in on the causes.

Multi-vari charts (Fig. 6.5) are used in stratified experiments to establish whether the variation is within a unit (position variation), from unit to unit (cycle variation), or from time period to time period (temporal).[8] A case history of full factorial experimental design is in Sec. 6.3.2.

One of the major strengths of the Shainin method is that it gives equal importance to the effects of interactions. Shainin reminds us of many engineering laws governed by interactions[5] such as $PV = MRT$, which is a three-factor interaction affecting the pressure and volume. The author has used the interaction of two variables (gas pressure and voltage) in solving the problems of seemingly mysterious corona failures in high voltage tables. The study of interactions is the main reason for using full factorial experiments. These may incur more cost, but the analysis is thorough. Also, it is easier for a practitioner to understand just one simple way of experimenting.

Taguchi approach.* The renewal in the use of design of experiments (DOE) methods in industry since 1982 is primarily a result of the work of Dr. Genichi Taguchi. Classical DOE focuses on identifying factors that affect the *level* (or *magnitude*) of a product/process response, examining the response surface, and forming the mathematical prediction model. Taguchi, however, recognizes that DOE should be applied in both product and process development to identify the factors that affect the *variability* of the response. The combination of different factor levels, identified from the DOE with significant effects on response variability and checked by a confirmation run later, is likely

---

*The author thanks C. Julius Wang of Chrysler Corp. for his generous contribution of material on the Taguchi approach.

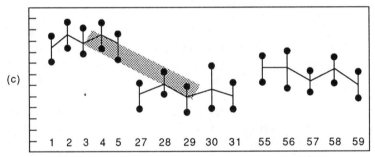

Figure 6.5   Multi-vari charts. (*a*) Position variation (within unit); (*b*) cycle variation (unit to unit); (*c*) temporal variation (time to time). *(Source: Keki R. Bhote, World Class Quality, American Management Association, New York, 1988.)*

to result in a product/process response insensitive to the change of operating settings, environmental conditions, and material degradation. This pioneering work not only improves the product/process quality, but also enhances the reliability, since the product/process response is consistent over time and different operating conditions.

Taguchi's robust design methodology[8] applies a fractional factorial design technique, which means more factors are studied with less experiments at a loss of some information. It makes use of standard matrices for experimental design setups, called *orthogonal arrays*. Depending on the number of variables and levels of experiments, an appropriate array is chosen. The methodology studies more than the traditional number of effects, takes advantage of nonlinear relations between response and control factors, employs the operating window concept (explained later in this section), and selects appropriate quality characteristics to achieve three objectives: *efficiency* in generating information to make decisions, *economy* in developing a product/process, and *consistency* in product/process performance over time and operating conditions.[9]

Role of orthogonal arrays and linear graphs. Orthogonal arrays are statistically designed experimental layouts. (*Note*: To develop robust designs, one can use any method of design of experiments. Using orthogonal arrays is one approach proposed by Taguchi.) In robust design, an orthogonal array is used to achieve additivity of main factor effects. In other words, interactions are practically not welcome. This feature makes the analysis subjective to some extent. Reproducibility from laboratory test results to downstream production and field use is a big concern.[9] By applying orthogonal array experiments and performing confirmation runs to verify the additivity prediction, experimenters can get a highly consistent product performance in a timely fashion at a lower cost. One example of an orthogonal array is called the $L_8(2^7)$ array, which means it can be assigned up to seven 2-level factors and that only eight experiments are required (full factorial design would require 128 experiments to study seven variables).

Linear graphs (Fig. 6.6) developed by Taguchi can be used to facil-

$$L_8(2^7)$$

	Col	A	B	C	D	E	F	Left open
		1	2	3	4	5	6	7
Row								
1		1	1	1	1	1	1	1
2		1	1	1	2	2	2	2
3		1	2	2	1	1	2	2
4		1	2	2	2	2	1	1
5		2	1	2	1	2	1	2
6		2	1	2	2	1	2	1
7		2	2	1	1	2	2	1
8		2	2	1	2	1	1	2

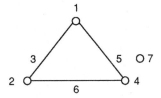

Figure 6.6 Example of orthogonal array and linear graph.

itate assignment of factors to each column. This figure also shows the $L_8(2^7)$ orthogonal array in which the factors $A, B, C, D, E,$ and F are assigned to six columns. Each column shows the levels of each variable in each of the eight experiments. Each row represents a specific experimental run. For example, row 2 indicates that the second experiment will have factors $A, B,$ and D at low level and the factors $E, C,$ and F at high level. Column 7 in this study is left open, since no factor is assigned to it.

The linear graph in Fig. 6.6 shows that each dot and line segment can be assigned a factor and the line segment also is the interaction of the two connecting dots. Therefore, column 3 will be a compound effect of interaction between the factors represented by columns 1 and 2— that is, factors A and D, factor B (the main effect) and some other interactions. This is known as *compounding*. In Taguchi's robust design, seeking main effects is more important than identifying interactions, and this feature makes the Taguchi approach somewhat subjective. The compounding effect, however, makes it possible to explore the significant main effects and to verify them in a confirmation run.[10]

Nonlinearity application. The advantage of using the nonlinear relation between response and control is best illustrated in Fig. 6.7. Sup-

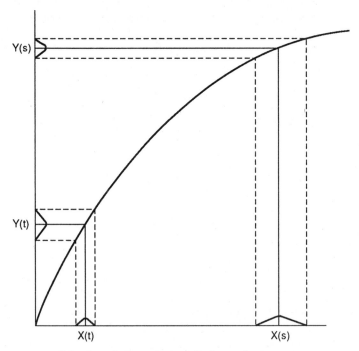

Figure 6.7 Selection of tolerances to optimize performance.

pose the product response (e.g., output voltage) target is $Y(t)$ and the corresponding factor (e.g., resistance) setting under the influence of all other factors should be $X(t)$. The resistance value has a distribution around $X(t)$ because of material inconsistency and manufacturing process variation. The resulting response voltage will have a large variation around the target. However, if the resistance value is changed from $X(t)$ to $X(s)$, the resulting response voltage $Y(s)$ has a more centered distribution, even though the resistance shows more variation than before. Since the response is now off target, adjustment factors identified from DOE—factors having strong impact on the level of response but limited impact on the variability—could be used to adjust the response back to the intended target $Y(t)$. This is sometimes called *two-step optimization.*[11–13]

Outer array application and parameter design. Some variables cannot be controlled or are impractical to control without a major investment. For example: eccentricity on an old machine or room temperature and humidity. But their effect on quality may be significant. Such factors are treated as noise but their effect is evaluated by means of an additional experimental setup called an *outer array.* The main orthogonal array is the *inner array.* Figure 6.8 illustrates the experimental layout for handling the outer array. This method is used in parameter design, which is the most critical step in Taguchi's three-step development strategy of *system design* (specifying the system requirements), *parameter design* (optimizing factor levels to achieve insensitivity with the least expensive components), and *tolerance design* (upgrading component tolerances when necessary to reach the desired target with the least extra cost).

Operating window concept. Taguchi proposes an operating window concept in which the larger the window, the more immune the process is to the production variation. An example[14] from a study of copying ma-

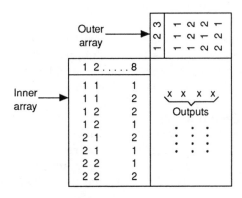

Figure 6.8 A conceptual layout of inner and outer arrays.

chine paper feeders illustrates the concept. Two failure modes, misfeed (no paper) and multifeed (too many papers) were of concern. The paper is fed by using the friction between pressing rollers and the paper, which is pressed by a spring. The spring force is critical, since at a threshold value F_1, the misfeed failure mode disappears and a single sheet feeds. Increasing the pressing force to a second threshold value F_2 introduces the multifeed failure mode. The objective then is to maximize the operating window formed by two threshold values (Fig. 6.9a). Another example is a steel alloy that achieves desirable ductility in a 25°F temperature range during a hot-rolling process. This temperature range is difficult to maintain. By adding silicon in the steel composition, the range was widened to 75°F (Fig. 6.9b).

6.2.4 Process FMECA

Design reviews are the best opportunity a quality assurance engineer has for voicing opinions and getting the appropriate action implemented. At this time, the hardware has not been made, or has money been invested in tooling. Therefore talk is still cheap and has its greatest impact on reducing life-cycle costs. Engineering changes are practically free.

The process failure-mode, effects, and criticality analysis should consider the performance of the entire manufacturing process, which is much more than just the equipment. It includes support services and operating procedures. The purpose is to use brainstorming to develop the best possible process. The product design engineer is very much a team member in the review on an as-needed basis. The process or the manufacturing engineer may or may not chair the review team but should assume technical leadership. For a process to work right, it must incorporate the expertise in variation control that comes from quality assurance engineers and the expertise in fabrication methods that comes from manufacturing engineers. The reliability and the design engineers should be invited to join in a review at important milestones in the project.

To have a best product, a company must have the best team. It is equivalent to saying that if a company has the best design, best manufacturing, and best quality assurance, then its product must be the best. If any of these elements is not the best, the product will not be a market leader.

One of the major reasons for the failure of the team effort between manufacturing and quality assurance is a lack of integration. Process FMECA allows this integration to take place early. It requires quality assurance and manufacturing engineers to work together and integrate their skills. It is a design review tool for analyzing a new pro-

Figure 6.9   Operating window examples. (a) Operating window for paper
feeding; (b) operating window for steel ductility.

cess. It includes analysis of receiving, handling, and storing materials
and tools. Whenever possible, suppliers should also participate.

The procedure is as follows:

Step 1:   Construct a process flow chart. Include in it all inputs to
the process such as the materials; their storage, handling, and
transportation; and the tooling. Standard operating and maintenance procedures should also be identified.

Step 2:   Using a process FMECA format similar to Fig. 6.10, perform the analysis below.

Step 3:   Write down each major process step in column 1.

Figure 6.10   Process FMECA.

Step 4:   Write down what can go wrong (failure mode) for each process step. For example, if a process step for an assembly is "Insert a lockwasher," some things that could go wrong are: "wrong washer," "cracked washer," and "operator forgets to insert washer."

Step 5:   Write the possible causes (failure mechanism) for the dis-

Page 1 of 10

Dalton, Fraser, Henley Assembly Plants

(7) Prepared By J. Ford — X6521 — Ford B&A

199X/Lion 4dr/Wagon

(9) FMEA Date (Orig.) 9X 09 01 (Rev.) 9X 09 01

9X 03 01

(6A) Key Production Date 9X 08 26 — Job #1 (22)

Current controls (16) (17)	D e t e c t i o n	R. P. N.	Recommended action(s) (18) (19)	Area/individual responsible & completion date (20)	Action results				
					Actions taken (21)	S e v e r i t y	O c c u r r e n c e	D e t e c t i o n	R. P. N.
Visual check each hour — 1/shift for film thickness (depth meter) and coverage	5	280	• Add positive depth stop to sprayer	Mfg engrg 9X 10 15	Stop added, sprayer checked on line	7	2	5	70
			• Automate spraying	Mfg engrg 9X 12 15	Rejected due to complexity of different doors on same line				
Test spray pattern at start-up and after idle periods, and preventative maintenance program to clean heads	3	105	• Use Design of Experiments (DOE) on viscosity vs. temperature vs. pressure	Mfg engrg 9X 10 01	Temp and press limits were determined and limit controls have been installed — control charts show process is in control $C_{pk} = 1.85$	7	1	3	21
Preventative maintenance programs to maintain head	2	28	None						
Operator instructions and lot sampling (10 doors/shift) to check for coverage of critical areas	7	392	• Install spray timer	Maintenance 9X 09 15	• Automatic spray timer installed — operator starts spray, timer controls shut-off — control charts show process is in control $C_{pk} = 2.05$	7	1	7	49

SAMPLE

crepancy. For the lockwasher example, the cause may be "human error."

Step 6: Identify the effect of the discrepancy on the end product. Example: "Missing lockwasher may prevent product from performing its critical mission."

Step 7:   Assign criticality. This can be assigned as a multiplication of severity, frequency, and undetectability ratings, each rated on a scale of 1 to 10. Each organization has to define these three terms and the associated ranking system. The following definitions should apply to most organizations.

Severity may be defined as the impact on safety and functional performance. Multiple-fatality accidents or a failure resulting in enormous financial loss may be rated as 10, while negligible damages will be given the numerical value 1. Sometimes the effects on the process are not severe, but a defect may cause unsafe conditions for the user. In such cases the effects on the user should also be considered.

Frequency can be judged on the probability of occurrence. If the event is certain to occur, it will be assigned a rating of 10. The impossible event will assume the lowest rating. Some companies substitute *probability of failure* for frequency.

Detection can be defined as how late in the process the defect can be caught. If it can never be caught before shipping, then the detection rating is the highest.

Figure 6.11 shows the relationship of these criticality measurements to other FMECA items. It may be noted that severity is a function of the ultimate effect on the customer.

Step 8:   If the equipment is not yet fabricated, perhaps the design can be changed to avoid the problems exposed in step 7. For already-built equipment, preventive action may be possible. In most cases, life-cycle costs are lower if the corrective action is taken in design.

As an example of the efficacy of correction in design, consider the failure mode in which an operator forgets to add a lockwasher before putting on a bolt. Inspecting the assembled equipment for the presence of the lockwasher would be monotonous, and inspectors might

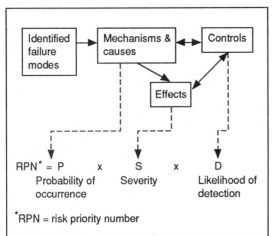

Figure 6.11   Relationship of criticality to FMECA items.

often think there is a lockwasher even if it is not there. A design approach would be to use an automated assembly fixture to assure that the lockwasher will always be inserted. An alternative design choice would be to make the washer an integral part of the bolt. A third choice would be to eliminate the need for the washer in the product.

General corrective actions, in order of precedence, are

1. Change the process so that the cause of the defect is eliminated permanently.

2. Install some kind of warning system so that the operator will know what is wrong with the process before it produces defective pieces.

3. If these actions are not effective, give the operator special training.

There may be a few hundred failure modes in an FMECA. Not every one requires a design action. In the interest of working effectively and efficiently, the important failures should be determined by evaluating the risk assessment results and handled first. The safety-related failures must receive top priority, even though their total rating may not be the highest. In cases where the corrective action is not obvious, a fault tree (Chap. 5) or process analysis maps (this chapter) may help in identifying root causes and actions.

6.2.5 Quality assurance plans for procurement and process control in the design phase

Some planning is required before going into production. These efforts will allow for a smooth transition from design to production.

Equipment procurement plans. Procurement plans should include verification of equipment performance with suppliers. Of course, performance of some large capital equipment cannot be verified until it is installed at the customer's site. In that case, performance at subsystem levels should be verified. Other activities may include testing for component interchangeability, statistical tolerance analysis, establishing machine capabilities, and pilot runs.

Process control plans. All the critical points where a major defect can be introduced should be identified before going into manufacturing. Quality function deployment and process FMECA should be applied to generate process control plans, since this is the only way to assure that the right things are controlled. Every control should be traceable to customer satisfaction, which is reflected in the QFD. Plans should include methods for controlling these critical process points, definitions of acceptance standards, and ways of monitoring all defects. The critical points should be highlighted in the manufacturing method sheets.

Component procurement quality plans. The plans for procured components should be in place before the pilot run starts. Once defective parts come in, they destroy any hope for a successful pilot run. The supplier and customer must work together because supplier components require as much quality as any other component in the product. At least four areas warrant concern and cooperation:

1. Process control implementation throughout the production lines

2. Component qualification

3. Standard and special screening tests (see Chap. 3)

4. Closed-loop failure management

6.3 Quality Assurance in the Manufacturing Phase

Quality assurance activities in manufacturing are aimed at assuring that processes are always up to standards and that manufacturing protocols for quality assurance are in place. The ownership of each procedure and process must be logically and clearly identified. For instance, if the material is discrepant, it should be the production foreman who has ownership of the problem—not the quality control engineer. It is the production foreman's job to ship only good products. The quality control function must go on, even if the quality control engineer is not available.

6.3.1 Evaluation of pilot run

If the manufacturing yield is going to be low, the pilot run is the place to correct it. In no way should the customer relationship be injured by the shipment of shoddy products under pressure of schedules. With a lot of "homework" done in advance, the pilot run yield should be high.

Preparation activities should also concentrate on assuring that process engineers have written process specifications to establish process performance standards. The operating procedures should also be documented to establish human performance standards. The yield depends on both.

Usually the process standards are poorly defined and often are not defined at all. A solder equipment operator is told to produce a good solder joint but is rarely told the best temperature range for the solder. Similarly, rarely does a manufacturing engineer specify when to change bearings, filters, and lubricating oils. They are usually changed after the process starts to go out of control. In over 95 percent of cases, equipment operators have poor process specification. A typical statement is "Heat the part to 200°C." This statement is completely inadequate. It does not tell the operator if 190°C is acceptable or how closely the variation must be controlled.

Operating procedures are just as important. In the above example, the operator should also be instructed what to do if the temperature is above or below certain values. Other information in the operating procedures should include when to add lubricant, how much lubricant, when to stop the process, how to monitor the process, and how to prevent production of defective units.

Operating procedures also include preventive maintenance procedures. The maintenance procedures should be based on knowledge of the process. They should be designed to prevent manufacture of defective parts. The belts, bearings, and fans should be changed before they have negative effects on quality. In other words, the maintenance personnel must be familiar with statistical process control techniques. In one case, a chemical concentration in a solution was to be controlled closely but the process was always out of control. The reason was that one of the chemicals came in 5-gallon drums and the operator poured all 5 gallons, even though only 1 gallon was required. The operator was never told how to determine the right amount to pour. Without good operating procedures, a process will never be stable, and statistical controls will be meaningless.

In any way possible, quality function deployment, discussed in Sec. 6.2.1, should be used to generate the manufacturing process performance standards and operating procedures (for operation standards). Thus, customer needs are converted to measurable (controlable) engineering characteristics, these engineering characteristics are then deployed to different parts specifications, and manufacturing process performance standards and operating procedures are finally developed from the associated parts specifications. This exercise guarantees that what the manufacturing process is doing really relates to the original customer expectations and anything to be changed will directly affect the customer wants.

6.3.2 Process control

It cannot be emphasized enough that someone in the organization should have ownership of the process specifications and operating standards. Every time a defect is produced, the validity of these documents should be questioned. This protocol should be well understood.

Some continuous improvement activities during the manufacturing stage may be validation of interchangeability, experiments to improve relationship of tolerance to product performance, verifying machine capabilities, establishing accuracy and precision of quality assurance measuring devices, improving preventive maintenance, improving operator training, and conducting audits. How well these activities are performed will determine whether the manufacturing process can be controlled.

Process control tools are used to get the best yield out of the existing equipment. Nonetheless, if a new item of equipment is needed, a tradeoff analysis should be made. Operating personnel are trained to use process control techniques which may or may not involve statistical control charts. Many controls can be put in place without control charts. The best strategy is to eliminate the source of problems and avoid costly controls.

Many try to use control charts on a characteristic that is out of specification without proper planning. They take months to find out that the process cannot be controlled without controlling the causes. They waste time and money. Unfortunately, most companies belong in this category. Proper planning requires the use of analysis tools. Fishbone diagrams, design of experiments, and process analysis maps are useful problem-solving tools and prevention tools to improve the process. Lots of people believe the myth that putting statistical process control (SPC) charts in place and doing some fishbone diagrams and probably limited DOE will bring their process back to life without even noticing that they have an inferior process design at the outset.

Identifying causes of variation. Traditionally, so-called Ishikawa diagrams (fishbone diagrams) such as that in Fig. 6.12 are constructed to identify sources of variations. These variations can come from the "six M's": man, materials, measurements, methods, machines, and Mother Nature (environmental factors).

Fault-tree analysis is another approach. The author used a simplified fault-tree approach while working with McElrath & Associates, management consultants, at one of the largest steel plants. Since 1981 this tool has been used at several corporations for statistical process control training with dramatic results. In one case, the cost of a component was reduced by about 50 percent. The tool offers flexibility and ease of construction.

Fault-tree analysis can be simplified as a process analysis map (PAM), Fig. 6.13. Each level in a PAM identifies the causes of varia-

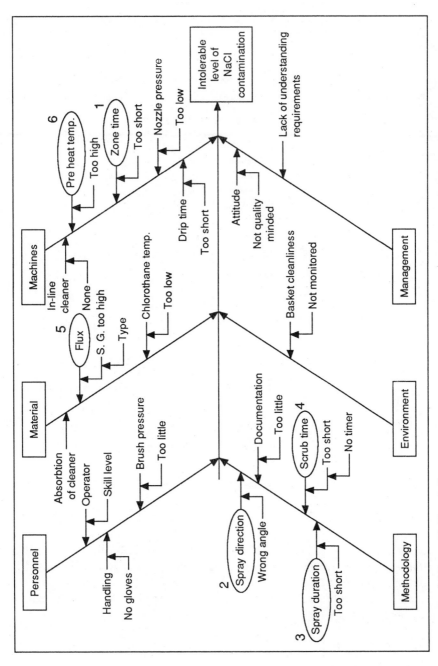

Figure 6.12   Ishikawa (fishbone) diagram.

Figure 6.13   Example of a process analysis map (PAM).

tions at the previous level. Usually three levels are sufficient. If it is necessary to go to more levels later, a sheet of paper can be added to extend the chart.

At each level, the causes attributable to the six M's are identified. Typical causes are

Man: Machine operator, maintenance technician, supervisor.

Machine: Production equipment, tools, instruments, gages.

Measurement: Production equipment accuracy and measuring instrument accuracy and precision.

Methods: Machine control methods, standard operating procedures, maintenance procedures, repair procedures.

Materials: Supplier parts, fabrication materials such as epoxies, fluxes, welding rods.

Mother Nature: Humidity, temperature, dust, chemicals, contamination, radiation.

To manufacture products that are cheaper and of higher quality at the same time, one should follow the famous 20/80 principle which says that about 20 percent of causes contribute to 80 percent of losses. The percent contribution of each cause to the problem is therefore estimated so that process controls are placed at the critical points. The PAM of Fig. 6.13 identifies more than 30 causes for the variation in the thickness of a material. It would be costly to run statistical process controls on all causes. Some causes can be technically detrimental but may have very little effect on the process. Therefore, one must verify the contribution of the critical causes by design of experiments. A good strategy is to use judgment in selecting three or four causes and experimentally test their influence. Then identify other important variables and check them the same way.

Verifying the influence of causes. The next step is to run experiments to isolate the critical causes in the PAM. Methods proposed in earlier references and in Ref. 15 are among those commonly used. A technique presented below (the author learned to appreciate this full factorial technique from Shainin) is reasonably economical and will work most of the time. If one still cannot identify most of the critical causes, then more in-depth and more efficient studies and experiments can be selected from the references.

Decide which three or four variables are most critical. Temperature, eccentricity, and roll condition were picked from the PAM in Fig. 6.13. Call them variables A, B, and C. To study the influence of a variable, perform experiments at at least two levels (high and low) for each variable. To assure the experiment is performed for all the possible combinations, perform $2^3 = 8$ experiments, where 2 represents the number of levels and 3 represents the number of variables.

The experimental design matrix is constructed by using the principles of full factorial experiments as shown in Fig. 6.14. The + and − signs are equivalent to levels 1 and 2 in the Taguchi method and represent the high and low levels of the cause. In this

	Factor A	Factor B	Factor C	AB	BC	AC	Yield
1	+	+	+	+	+	+	75
2	+	+	−	+	−	−	29
3	+	−	+	−	−	+	25
4	+	−	−	−	+	−	7
5	−	+	+	−	+	−	48
6	−	+	−	−	−	+	56
7	−	−	+	+	−	−	39
8	−	−	−	+	+	+	4

Figure 6.14   Experimental design matrix. (*Note*: Interaction ABC is not shown but can be assessed.)

method, half the experiments for the first variable A are assigned at high (+) level and the other half at low (−) level (see column labeled Factor A). The set of four continuous +signs under A is called a *run*, as is the continuous set of four negative signs. To assign levels for variable B, use an equal number of + and − signs for each run of variable A (see column labeled Factor B). Similarly, for variable C, make half the experiments at high level and half at low level for each run of variable B.

In these experiments, one needs to know the effects of interactions of two or more variables. No additional experiments are required because the interaction effects can be assessed from the same data. The columns AB, BC, and AC show the influence of these interactions as follows: To construct column AC, multiply the signs of variable A and variable C to determine the sign under variable AC; two positive or two negative signs multiplied yield a + sign, and two dissimilar signs multiplied yield a − sign. As shown below, the resulting signs will be used in determining the influence of the interaction.

The experimental design in Fig. 6.14 shows that the first experiment will be performed with all the three variables at high level. The second experiment will have variable A at high level, variable B at high level, and variable C at low level. There are altogether eight experiments, which include all the possible combinations of A, B, and C.

The last column shows the yield (percent product in the target range) for each experiment. Any other measurement of inherent process quality would be just as appropriate. Looking at the data one can say that the highest yield is obtained when A, B, and C are at high level, but such conditions may be too expensive to maintain all the time. Also a hidden variable may be responsible for some of the yield. Scientifically it is better to establish the influence of the individual variable and maintain control on it all the time. Even better, eliminate the source of variation. In this case the source of variation, eccentricity, could be eliminated only by a major equipment change, and management elected to use statistical process control instead.

To determine the effect of the variable as other factors vary, one may compare the average yields when the variable is at high level and at low level. This analysis is shown in Fig. 6.15. For variable A, the average yield for the first four experiments (A at high level) is 34 percent and the average yield for the last four experiments (A at low level) is 36 percent. Since the difference in the yield is only 2 percent, the effect of variable A does not seem to be significant. Variable B, on the other hand, seems to have a considerable effect. The average yield for B at high level (experiments 1, 2, 5, 6) is 52 percent versus 18.8 percent at low level (experiments 3, 4, 7, 8). The influence of interactions is assessed in the same manner. If we wish to know the effect of AC, we can compare averages when interaction AC is at high level (experiments 1, 3, 6, 8) and when interaction AC is at low level (experiments 2, 4, 5, 7). The corresponding yields, 40.0 percent and 30.8 percent, show that the interaction seems to influence the yield. If the reader is interested in accuracy, then Analysis of Variance (ANOVA) and other techniques found in the references may be used.

The data in these experiments show the variables B and C have a

Factor	Avg. yield at high level	Avg. yield at low level	Difference
A	34	36	−2
B	52	18.8	33.2
C	46.8	24	22.8
AB	36.8	34	2.8
BC	33.5	37.3	−3.8
AC	40.0	30.8	9.2

Figure 6.15   Comparison of yields.

Figure 6.16   Reversal in yield with control of two variables.

significant effect on yield. Often the yield will improve dramatically when just a few variables are controlled. In this case, the statistical process control of only two variables—*B* and *C*—produced a complete reversal in yield (Fig. 6.16).

In summary, design of experiments is a powerful tool, and tools such as PAM and fishbone diagrams are just as necessary. The key is to eliminate sources of variation, if possible. Short of that, variability must be reduced first, then the mean must be controlled close to the target value.

Statistical process control.   Ideally a cause of excess variation should be eliminated by the process engineer in the equipment design. Many times this can be done, but companies normally do not invest enough in this area. For an inherently unreliable process, the control chart may be necessary. The purpose of control charts is to study the variation in the process and monitor its stability. Actually, control charts *do not* control the causes of variation. Control charts *detect* a signal out of control, and other tools are needed to bring the process back into control. Also, once control charts detect a signal of out of control, one has to verify whether the causes are sporadic or systematic. A sporadic cause (such as the operator not replacing the coolant in the machine) can be eliminated very quickly. However, the removal of a systematic cause often needs a management decision. For example, a machine simply may not be able to hold the required tolerance because it is not designed for that accuracy. Then a management decision may be made to change the machine.

The generic steps in controlling the process are

Step 1:   Check whether the process is stable by constructing a control chart. If the process average or the standard deviation is not stable over a reasonable length of time, then some uncontrolled causes must be removed before the process can be controlled.

Step 2:   Determine the inherent process capability which may be computed at ±4 standard deviations from the data. To guard against production of nonconforming parts, a customer often specifies a so-called capability index C_p = 1.33. C_p is the ratio of engineering tolerance to the traditional 6 standard deviation process spread. For an area less than 100 ppm outside the specification, this ratio should be at least 1.33. (For detailed discussions of noncentered processes, see Sec. 6.3.3.) If the process capability range is wider than the engineering specifications, then some yield loss is expected, besides extra inspection cost. Therefore, an attempt must be made to remove or control some causes of variation by using information from cause-effect diagrams such as PAM or fault trees.

Step 3:   Keep removing or controlling the causes of variations and verify their effects on control charts until the C_p and ppm goals are met. The ppm goals will determine the maximum allowable size of the standard deviation.

Step 4:   Monitor control charts and train operators to recognize the signals for processes going out of control. They should be able to correct the condition before any defective piece is produced. Control charts typically give warning of the process going out of control. Many books on statistical quality control cover such topics.[16]

There are two types of basic control charts: (1) control charts for variables, such as X-bar and range charts, and (2) control charts for attributes, such as p charts and c charts. Modified charts are used for special applications such as precontrol charts, moving average charts, X-bar and S charts, and median and S charts. Most charts are described in Ref. 17.

Control charts for variables.   These charts are used when the variable can be measured, such as a dimension on a shaft. The most popular charts are X-bar charts for controlling averages and R charts for controlling range. (Range has to be controlled because it is a measure of piece-to-piece variation. It shows whether a variation is normal or abnormal.) Figure 6.17 shows average (X-bar) and range (R) charts for a chemical solution where pH value must be controlled. At predeter-

Figure 6.17 Control chart for average and range.

mined time intervals, depending on how stable the process is (in this case twice every shift), a small sample called a subgroup is checked. Usually a sample subgroup of five is chosen (in this case the subgroup contained three readings). Over a longer period, sufficient data are collected so that statistically sound judgment can be made about process capability. The process limits are determined as follows:

Process average = $\overline{X}$ = average of all the subgroup averages

$$= \frac{\Sigma \overline{x}_i}{n} = 6.233$$

Average range = R = average of all the ranges

$$= 0.563 \text{ from the data}$$

To compute limits of process variations ± 3 standard deviations, the following rules can be followed, with constants such as A_2, D_4, and D_3 provided in Table B in the appendix.

The control limits for the X-bar chart are

$$UCL = \overline{X} + A_2\overline{R}$$

$$= 6.233 + 1.023 \times 0.563$$

$$= 6.809$$

$$LCL = \overline{X} - A_2\overline{R}$$

$$= 6.233 - 1.023 \times 0.563$$

$$= 5.657$$

where UCL is the upper control limit and LCL is the lower control limit.

The control limits for the R chart are

$$UCL = D_4\overline{R}$$

$$= 2.574 \times 0.563$$

$$= 1.449$$

$$LCl = D_3\overline{R}$$

$$= 0 \times 0.563$$

$$= 0$$

Caution: These are the limits on averages of the subgroups made by the process, not on the individual pieces. For individual piece variation see Ref. 17.

Control charts for attribute data. Many times no measurements are taken on the variable. The quality levels are defined in terms of fraction of production nonconforming or number of nonconformities per unit of a finished product. The control charts for these applications are called p chart and c chart respectively. The control principles are the same as for the charts for variables. Below are the computation rules.

The control limits for p charts are

$$UCL = \bar{p} + 3 \sqrt{\frac{\bar{p}(1 - \bar{p})}{n}}$$

where p = average percent defective.

$$LCL = \bar{p} - 3 \sqrt{\frac{\bar{p}(1 - \bar{p})}{n}}$$

$$UCL = \bar{c} + 3\sqrt{\bar{c}}$$

where c = average defects per unit.

$$LCL = \bar{c} - 3\sqrt{\bar{c}}$$

Constructing control charts for the sake of records is meaningless if the inherent process capability is not improved. The next step is to reduce the variation and adjust the mean of the process to shift it as close as possible to the specified target, before any nonconforming parts are fabricated.

The control charts give adequate warning of the process going out of control. For a stable process, the data points should be equally and randomly distributed around the average. The chance of a point falling outside the control limits is 3 in 1000. The chance of two points outside the control limits is about 9 in 1,000,000. When this happens, one can be almost certain that the process is no longer acceptable. Similarly, when seven or more points in a row show an upward or downward trend, the process is going out of control.

6.3.3 PPM control for world-class quality

Traditional inspection sampling plans like Mil-Std-105[18] are no longer appropriate. Sample size in thousands is required when one or more nonconformities are acceptable. For 100 ppm quality (0.01 percent acceptable quality level), the plan calls for 100 percent inspection up to a lot size of 1250, and no defects are permitted. If the process is producing more than 100 ppm defects, then practically every lot is going to be rejected. Even if the process produces world-class quality (100 ppm or less), about 12 percent of the lots are questionable, as shown by the following computation.

Suppose a batch has a quality level of 100 ppm, or 1 defect per 10,000. The probability of a good piece then is .9999. The chance that all pieces are within specification is (.9999)1250 = 1249.875. In other words, 11.75 of the lots are expected to have 1 or more pieces defective. Experience tells us that if an inspector is checking a sample of 1250 pieces and if the defects are rare, the inspector is likely not to notice the defects. The bad lots are expected to go to the customer. On the other hand, if the bad lots are caught, then entire production is suspect, resulting in 100 percent inspection. Regardless of which way the results go, the cost of inspection is too high. A better alternative is to narrow the process variation and monitor the stability of the mean at the specified target. There is a tradeoff between cost of inspection versus cost of engineering. Investing in engineering is generally cheaper if the root causes of variation are controlled (preferably eliminated).

Global competition, requiring higher quality at lower costs, has forced international companies to do the right things at the right time. As a result, the informal standard of nonconforming levels of less than 100 ppm has emerged. As mentioned earlier, world-class quality level is a moving target set by the best producer in the world. For a complex product and emerging technologies it may not be possible to achieve this standard. Such products may have a ppm as high as 15,000. The moral is that whoever comes into the global market first with low price and high quality sets the standard for others.

The criteria for producing less than 100 ppm nonconformance can be explained with the help of Fig. 6.18. Since a process is rarely exactly at the center of the engineering specification, analyze the worst side, which will produce more nonconformance. (If the process is centered,

Figure 6.18 Process capability and ppm level estimation.

either side will do.) In this case, side AB is the worst case. Table A in the appendix shows that for $Z = 4\sigma$, the nonconformance rate is 30 ppm, which meets the requirements of a maximum of 100 ppm. Therefore, if the distance AB is at least equal to 4 standard deviations and if the process is stable, then the process will always produce parts within the 100 ppm level. This criterion is explained in a slightly different way in industry.[16] It is written as

$$C_{pk} = \frac{Z}{3} = 1.33 \quad or \quad Z \geq 4.0$$

where C_{pk} is the capability index and Z is given by the expression

$$Z = \frac{|\text{specification} - \text{process average}|}{\text{standard deviation}}$$

if the process average is somewhere within the engineering specifications (see Chap. 2).

Example 6.3 A process is required to maintain a value within 150 and 225 lb/in². Calculate (a) the nonconforming ppm level produced by this process. (b) What should be the standard deviation for less than 100 ppm nonconformance?

Solution (a) The nonconformance for side AB of Fig. 6.18 is

$$Z_{AB} = \frac{225 - 185.2}{17.9} = 2.22$$

From Table A in the appendix, the fraction nonconformance is 1.32. The nonconformance for side AC is

$$Z_{AC} = \frac{185.2 - 150}{17.9} = 1.97$$

From Table A, the fraction nonconformance is 2.44. The total nonconformance is

$$Z_{CB} = 1.32 + 2.44 = 3.76 = 37,600 \text{ ppm}$$

(b) The worst side is AB. The criterion is $Z_1 \geq 4.0$. That is,

$$\frac{225 - 185.2}{\text{Standard deviation}} \geq 4.0$$

From this equation, the standard deviation should be less than $(225 - 185.2)/4$ or 9.95.

It has been mentioned that, to avoid inspection, a process must be stable. If the average keeps moving, controlling the standard deviation may not help. The secret of controlling the average is to make sure the maintenance and operating procedures are sound and well-understood. It is usually the preventive maintenance people who really have control over the equipment.

Complete control on a process requires monitoring seven parameters: product upper specification limit (USL), product lower specification limit (LSL), product target T, process mean M (central tendency), process standard deviation (variation), process skewness (a measure of symmetry), and process kurtosis (a measure of peakedness), in most applications. However, C_{pk} on the previous page considers only four of the seven parameters. Thus, the C_{pk} value is not unique; two processes can have the same C_{pk} value but different distributions. For example, consider two processes with $T = 15$, USL $= 20$, and LSL $= 10$. Process A has $\mu_A = 15$, $\sigma_A = 1.25$ and process B has $\mu_B = 18$, $\sigma_B = 0.50$. Both processes have $C_{pk} = 1.33$ but are totally different processes.

6.3.4 Working with suppliers

One hundred percent inspection is costly for the buyer as well for the supplier. They must work as a team to reduce the cost of inspection and increase the yield. This will lower the costs of components. The buyer must give a clear indication of what quality characteristics are critical. Simply asking suppliers to use statistical process control will raise costs. A very large automobile company sent such a directive to its suppliers. One of the suppliers started to put SPC on hundreds of dimensions, even though on 80 percent of them it did not make any difference whether they were in control or not. The result was that too much money was wasted on controlling dimensions that did not matter and not enough effort was put on critical dimensions. Some suppliers were controlling the dimensions but not their causes. They could never keep the processes stable and always fell back to 100 percent inspection.

One of the best tools is to have proper communication for tracking continuous improvements in process capabilities. The savings resulting from improved yield and less inspection should be shared by the buyer and the supplier. Eventually the ppm goals will be met by the supplier, and the buyer no longer needs to monitor. This is the most effective tool for a just-in-time system where the supplier ships parts straight to the production floor of the buyer. Since no warehouse inventory is required, the savings are enormous.

Caution: Buyers of components should have the same standard of excellence for their own final assemblies. Frequently a buyer will demand perfection from the supplier but its own final product is shoddy.

6.3.5 PPM assessment

Quality has become a prime factor in procurement. Buyers want to know the ppm level of components. Several organizations are involved in writing standards for ppm levels. One such standard is the ANSI/

EIA-555 Lot Acceptance Procedure for Verifying Compliance with the Specified Quality Levels in PPM (1989), developed by the Electronic Industries Association, 2000 I Street, Washington, DC 20006, and published in conjunction with the American National Standards Institute, 1430 Broadway, New York, NY 10018. It classifies nonconforming devices and components into five categories.

PPM-1: Functional nonconformances which are inoperative.

PPM-2: Electrical nonconformances which fail to meet specified electrical parameters.

PPM-3: Visual/mechanical nonconformances which fail to meet specified visual/mechanical characteristics.

PPM-4: Hermetic nonconformances which fail to meet hermetic requirements of a product.

PPM-5: All nonconformances including PPM-2, PPM-3, PPM-4 and all other specification nonconformances, excluding administrative requirements.

This standard contains two methods to estimate the ppm level. Method A is the worst-case estimate and is covered below. Method B is based on lot acceptance rate and can be found in the standard. The worst-case estimate is given by

$$\text{ppm in any class} = \frac{0.7 + \text{total nonconforming}}{\text{total number inspected}} \times 10^6$$

The number 0.7 represents a minimum number of nonconforming devices expected with roughly 50 percent confidence when a sample of n items is taken from an infinitely large population.

Example 6.4 Assess the ppm level when 3000 components are inspected and no nonconforming devices are found.

Solution

$$\text{ppm} = \frac{0.7 + 0}{3000} \times 10^6$$

$$= 233$$

Example 6.5 A manufacturer wants to develop a sampling plan which will allow less than 100 ppm nonconforming devices. Compute the sample size required when (1) no nonconforming devices are acceptable and (2) one nonconforming device is acceptable.

Solution (1) For acceptance number equal to 0,

$$\text{ppm} = 100 = \frac{0.7 + 0}{N} \times 10^6$$

Therefore,

$$N = \frac{0.7}{100} \times 10^6 = 7000$$

(2) For acceptance number equal to 1,

$$\text{ppm} = 100 = \frac{0.7 + 1}{N} \times 10^6$$

Therefore,

$$N = \frac{1.7}{100} \times 10^6 = 17,000$$

6.4   Quality Assurance in the Test Phase

In a large job-shop system such as a nuclear power generator, the test stage has a very critical influence on the schedule. Every precaution should be taken to design and manufacture the system right so that the system will pass the acceptance test in the least amount of time. The same concept applies to mass-produced products such as washing machines, coffee makers, and automobiles, where the test cycle is relatively short but the financial risk of rejecting the entire production is very large. In determining the adequacy of the test, a closed-loop system between field failures and the test program is necessary.

6.4.1   Qualification versus production testing

Qualification tests are those applied to prototypes and initial production. These tests are made rigorous to prove the product will work during its intended life. The tests are usually destructive. Production tests are much less rigorous. Their usual purpose is to make sure the product meets functional requirements.

The biggest mistake quality control departments make is that they rarely perform complete qualification tests once the device is in production. The author was once asked to investigate the cause of "wet bikes" going into the sea and never coming back. At least 11 persons died when they took these motor-cycle-like watercraft into the sea for recreational purposes. The vehicles are designed to be driven on water, but unfortunately a relay absorbed salt water and short-circuited the electrical system. During early design, the relays were leak-tested under the toughest environmental conditions with highly accurate leak-testing equipment. However, the production units were not subjected to such tests as thermal shock and accelerated aging, with the result that marginal units managed to go to the customers. If the company had done qualification tests periodically, these defective relays would have been caught in the factory. The company is out of business now. The best guide for ensuring a good system is to qualify several production units at a predetermined interval. The frequency should depend on how good the process controls are. This is the best insurance against production errors.

6.4.2 Industry standards

A word to the wise. Don't use industry standards as the only standard. These usually set a minimal level of performance. Every product usually requires unique tests which can be determined from analyses such as failure-mode, effects and criticality analysis, fault-tree analysis, and preliminary hazard analysis.

6.5 Quality Assurance in the Use Phase

For an existing product, quality assurance should continue in the use phase. Most companies instead concentrate on reducing costs in the manufacturing phase. Meanwhile many customers are turning away because of substandard characteristics unrelated to manufacturing. For example, car buyers may not like many small buttons and complicated instructions on their radios or they may prefer a fuel injection system instead of a carburetor system. Therefore, quality assurance must survey the real concerns of customers. This information should be fed back to product and process engineering to make improvements, to update the QFD results, and to ensure that similar mistakes are not made again. That's where the secret of quality control lies!

6.6 Topics for Student Projects and Theses

1 Critique the role of the manufacturing engineer in controlling the manufacturing processes. Develop a model for manufacturing engineering for assuring high quality at lower costs.

2 Most companies are unable to control quality because they fail to control operating and maintenance procedures. Develop a quality control model highlighting the role of these procedures.

3 Most practitioners complain that their management's commitment to quality has been ceremonial rather than real. Develop a model for top management involvement in quality assurance.

References

1. *Total Quality Management: A Guide for Implementation*, DoD 5000.51-6 (draft), U.S. Department of Defense, Washington, D.C., March 23, 1989.
2. Harold E. Vannoy and James A. Davis, "Test Development Using the QFD Approach," Technical Paper 89080, SAE, Warrandale, Pa., 1989.
3. John R. Hauser and Don Clausing, "The House of Quality," *Harvard Business Review*, May-June 1988.
4. Raghu N. Kackar, "Taguchi's Quality Philosophy: Analysis and Commentary," *Quality Progress*, 1986.
5. Dorian Shainin, "Better than Taguchi Orthogonal Tables," *ASQC Quality Congress Transactions*, 1986.
6. Genichi Taguchi, *System of Experimental Design*, vols. 1 and 2, Don Clausing (ed.), Unipub–Kraus International Publications, White Plains, N.Y., 1987.
7. Charles R. Hicks, *Fundamental Concepts in the Design of Experiments*, McGraw-Hill, New York, 1973.
8. Keki R. Bhote, *World Class Quality*, American Management Association, New York, 1988.
9. Genichi Taguchi, "Development of Quality Engineering," *ASI Journal*, vol. 1, no. 1, 1988.
10. C. Julius Wang, *Taguchi's Robust Design Methodology in Quality Engineering*, unpublished training booklet, Chrysler Corp., 1989.
11. Madhav S. Phadke, *Quality Engineering Using Robust Design*, Prentice-Hall, Englewood Cliffs, N.J., 1989.
12. R. V. Leon, A. C. Shoemaker, and R. N. Kackar, "Performance Measures Independent of Adjustment," *Technometrics*, vol. 29, no. 3, 1987.
13. G. Taguchi and M. S. Phadke, "Quality Engineering Through Design Optimization," *Conference Record*, IEEE GLOBECOM '84 Meeting, IEEE Communications Society, Atlanta, November 1984.
14. Yuin Wu, "Reliability Study Using Taguchi Methods," *ASI Journal*, vol. 2, no. 1, 1989.
15. G. E. P. Box, W. P. Hunter, and J. S. Hunter, *Statistics for Experimenters: An Introduction to Design, Data Analysis, and Model Building*, John Wiley & Sons, New York, 1978.
16. *Continuing Process Control and Process Capability Improvement*, Ford Motor Co., Dearborn, Mich., 1985.
17. J. M. Juran and F. M. Gryna (eds.), *Quality Control Handbook*, McGraw-Hill, New York, 1988.
18. *Sampling Procedures and Tables for Inspection by Attributes*, Mil-Std-105, Naval Publications and Forms Center, 5801 Tabor Avenue, Philadelphia, PA 19120.

Further Reading

Anderson, V. L., and R. A. McLean, *Design of Experiments*, Marcel-Dekker, New York, 1974.
Besterfield, D. H., *Quality Control: A Practical Approach*, Prentice-Hall, Englewood Cliffs, N.J., 1979.
Continuing Process Control and Process Capability Improvement, Ford Motor Co., Dearborn, Mich., 1985.
Control Chart Methods of Analyzing Data, ASQC Standard B2/ANSI 21.2, American National Standards Institute, 1430 Broadway, New York, NY 10018.
Deming, W. Edwards, *Out of the Crisis*, Massachusetts Institute of Technology, Center for Advanced Engineering Study, Cambridge, Mass., 1986.
Diamond, W. J., *Practical Experiment Designs*, Van Nostrand Reinhold, New York, 1981.

Feigenbaum, A. V., *Total Quality Control*, 3d ed., McGraw-Hill, New York, 1983.

Grant, E. L., and R. S. Leavenworth, *Statistical Quality Control*, 5th ed., McGraw-Hill, New York, 1980.

Guide for Quality Control Charts, ASQC Standard B1/ANSI 21.2, American National Standards Institute, 1430 Broadway, New York, NY 10018.

Harry, M. J., *Handbook for Achieving Quality Excellence: The Strategy, Tactics, Tools*, 1986.

Hicks, C. R., *Fundamental Concepts in the Design of Experiments*, McGraw-Hill, New York, 1973.

Ishikawa, Kaoru, *Guide to Quality Control*, rev. ed., Asian Productivity Organization, 1976.

Juran, J. M., *Juran on Planning for Quality*, ASQC Quality Press, Milwaukee, 1988.

Juran, J. M., and Frank M. Gyrna, Jr., *Quality Planning and Analysis*, McGraw-Hill, New York, 1970.

Kearns, D. T., "Chasing a Moving Target," *Quality Progress*, October 1989, pp. 29–31.

Ostle, B., and L. C. Malone, *Statistics in Research*, 4th ed., Iowa State University Press, Ames, Iowa, 1988.

Ott, Ellis R., *An Introduction to Statistical Methods and Data Analysis*, Duxbury Press, North Scituate, Mass., 1977.

Process Quality Management and Improvement Guidelines, Code No. 500-049, AT&T, Indianapolis, 1988.

Quality System Requirements, MIL-Q-9858.

Sampling Procedures and Tables of Inspection for Variables for Percent Defective, Mil-Std-414.

Scherkenbach, William W., *The Deming Route to Quality and Productivity, Roadmaps and Roadblocks*, Ceepress Books, Washington, D.C., 1986.

Schilling, Edward, *Acceptance Sampling in Quality Control*, Marcel-Dekker, New York, 1982.

Shoenberger, Richard, *Japanese Manufacturing Techniques*, The Free Press, New York, 1982.

Taguchi, Genichi, *Introduction to Quality Engineering*, Asian Productivity Organization, Unipub–Kraus International Publications, White Plains, N.Y., 1986.

Taylor, J. K. *Quality Assurance of Chemical Measurements*, Lewis Publishers, Michigan, 1988.

7

Logistics Support Engineering

7.1 Logistics Support Principles

A company will not be in business long if it does not support a product throughout its life. This support includes giving attention to the customer's need to keep the product operating. *Logistics engineering*, the branch of engineering that meets this customer need, bridges the gap between design engineering and product support functions. Design engineering decisions in the concept stage help to reduce logistics costs such as those of distribution, transportation, maintenance, spares, and documentation.[1]

To reduce support costs, engineers design systems for reliability and maintainability, but their success has been questionable. Lerner says:[2]

> In most widespread use is the reliability and maintainability audit. Here an independent group of auditors analyzes the designs of engineers and estimates such variables as mean times between repairs and cost. If these do not meet specifications, the engineers are told to modify their designs. This is where the process breaks down. Engineers do not as a rule know which part of the design needs improvement, and the design process is so far along that most often they wind up accepting less reliability and maintainability.

Logistics engineering involves all the support functions needed to preserve and maintain the integrity of a system once it has been shipped to the customer. Thus it is an engineering effort that minimizes costs and malfunctions during such activities as storage, handling, transporting, spares provisioning, repair, preventive maintenance, and training.

From a different angle, Coogan[3] explains:

> For simplicity's sake, we could say that every system is made of two major subsystems: the operating subsystem and the logistics subsystem. The

operating subsystem is the set of functions that perform the actual mission of the system. The logistics subsystem is the set of functions that are necessary to sustain the operating system.

The fact that logistic analysis is an engineering function means that it should be done in the engineering phase. When logistics engineering is done properly, the design of the entire system will be changed for the better.

7.2 Logistics Engineering during the Design Phase

Blanchard suggests the following measures of logistics effectiveness:[4]

Reliability factors (failure rates, component relationships)

Maintainability factors (downtime lengths, frequencies, costs)

Supply support factors (probability of spares availability, inventory considerations, mission completion considerations)

Test and support equipment factors

Organizational factors (direct and indirect labor time, personnel attrition, delays)

Facility factors (turnaround time, facility utilization, energy utilization)

Transportation and handling factors

Software factors (quality, reliability)

Availability factors (inherent, achieved, and operational)

Economic factors (inflation, cost growth, budgets)

Effectiveness factors (capacity, accuracy, and factors related to safety, reliability, maintainability and dependability, and life-cycle costs)

This list shows that logistic support analysis requires inputs from all assurance technologies to assess their effect on support costs. Many tradeoffs are required to optimize these costs. Support costs can be further reduced by investing in software such as that for fault isolation and automated testing.

7.2.1 Logistics specifications for existing products

Writing a good specification is the key to logistics engineering. It is here one can reduce life-cycle costs dramatically without a major investment in design changes.

For a new product, the logistics engineering analysis must be performed before the final specification is issued. Some inputs needed are FMECA, FTA, maintainability analysis, MTBFs, estimates of repair times, estimates of preventive maintenance workload, information related to shipping, storage and facilities, and training needs. This information is vital to improving the design. For example, if it is found that highly skilled electronic technicians are required for troubleshooting, it may be possible to redesign the equipment for technician transparency. That is, the troubleshooting can be simplified by a standard modular design so that the lowest-level technician can remove the failed module and replace it with a new one.

For a product which already has a history, Lerner[2] presents a technique used on the Lockheed C-130 tactical airlifter. The landing gear shown in Fig. 7.1 generated excessive unscheduled maintenance worker-hours. Analysis of data showed that key repairs were removal and replacement of the wheel-and-tire assembly and maintenance of the mechanisms that lower and retract the wheels. The technique is described in this case history from Ref. 2:

Repairing the wheel/tire assembly was complex, needing several types of equipment such as jacks, torque wrenches, and thread protectors oper-

Figure 7.1 Landing gear of a C-130 airliner.

ated by level-five specialists. The repair exposes hub bearings to dirt and grit.

A main reason for frequent tire failures was heat transfer from the brakes through the wheel to the tire. The main reason for retraction mechanism failure was friction from tight clearances. Finally, both systems had many similar components that were not interchangeable or standardized.

From these key deficiencies, the engineers generated SDTRs (supportability design-to requirements):

- Avoid close tolerances to raise reliability and diminish the skill required of maintenance personnel.
- Make parts interchangeable and reduce the total by 50%.
- Eliminate separate jacks with a self-jacking assembly on the landing gear.
- Do not allow wheel/tire removal to expose the hub or require specialized tools.
- Design wheel to keep heat transfer from the brakes to a minimum.

After the design engineers saw these concrete requirements, they came up with a much more supportable design. Rollers spring-loaded against their tracks replaced a close tolerance show-and-track guide mechanism. Jacking modules appeared in the lower part of the shock strut. Three components replaced the landing gear wheel: the hub, the torque cage with the brakes, and the wheel rim/tire assembly. The rim assembly is not in direct contact with the brake assembly, reducing heat flow, and perforations in the brake assembly admit cooling air. The modular design meant that each part could be repaired separately and with no special tools. Also, parts were standardized throughout.

Results were spectacular. Retrofits nearly tripled mean time between failures, reduced spare parts by 60%, shortened ground time for wheel/tire replacement fourfold, halved repair personnel, and lowered skill levels. Although the modular design added 600 lb to gear weight, longer lasting carbon brakes saved 400 lb, so performance was unaffected.

7.2.2   Logistics specifications for new products

Consider the chemical plant in Fig. 5.3, Chap. 5. Even from design sketches like this, many logistics concerns can be identified. Some such concerns will be

How can material handling equipment be placed underground? What material handling devices are required?

Is a crane needed?

Can all repairs be made underground?

Can all preventive maintenance be performed underground?

Are there any safety concerns during repairs?

What spare parts will be needed?

Can corrosion cause damage? What kind of quality assurance help is needed?

What information is needed in preventive-maintenance and repair manuals?

What kind of training personnel will be needed?

What kind of maintenance personnel will be needed?

Are all provisions considered for initial installation of the equipment?

What items must be operational at all times (minimum equipment list)?

Hundreds of questions such as these can be anticipated in early design. But, if special equipment is needed, it is better to design it at this stage. If some repairs would require enormously expensive spares, it is better to remove that failure mode by design or provide a backup at this stage. In fact, if very large and bulky material handling equipment is required, an underground design may be inefficient. Perhaps an on-the-ground facility with containment will be desirable. With advanced analysis and design changes, support costs can be dramatically reduced.

7.2.3 Design reviews

Logistics engineers often get involved in design reviews too late. They should participate in conceptual design reviews, identify all the support costs and concerns, and recommend design changes. A good goal for logistics engineering is suggested by the U.S. Air Force as "double R/half M," which means double the investment in reliability cuts maintenance costs in half. The increased investment in reliability is small compared to the giant reduction in maintenance costs. Principles of design reviews are covered in Chaps. 3 and 6.

7.2.4 Logistics support analysis

Logistics support analysis is a composite of systematic actions taken to identify, define, analyze, quantify, and process logistics support requirements. It evolves as the product development takes place. Maintenance analysis is a major element of this analysis.

For military systems, Mil-Std-1388[5] describes many tasks such as:

1. System-level logistics support analyses

    a. System design and operational concept influence
    b. Identification of supportability, cost, readiness drivers
    c. Support concept development
    d. Tradeoffs
2. Element-level logistics support analysis
    a. Support system optimization
    b. Resource requirements identification
    c. Task and skill analysis
    d. Early fielding analysis (to assess impact of introduction of the new product on existing systems)
    e. Postproduction support analysis
3. Supportability assessment and verification
    a. Supportability assessment plans and criteria
    b. Support concept verification
    c. Verification of resource requirements
    d. Identification and correction of logistics problems

7.2.5 FMECA for logistics support analysis

Maintenance, of course, is one of the major elements of logistics engineering. It is an essential tool for design analysis and maintenance planning. FMECA provides an analytical baseline for a variety of other purposes—for example, for reliability analysis (Chap. 3) and for manufacturing process analysis (Chap. 6). The procedure described below is for maintenance analysis.

    Figure 7.2 shows an FMECA model work sheet for maintenance planning.[6] Multiple use of FMECA in the same program requires careful planning to prevent duplication of effort and to properly interface with reliability, safety, and maintainability engineers. A separate work sheet is completed for each replaceable or repairable unit, called a *logistic support analysis candidate* (LSAC). A separate work sheet is also required for each structurally significant item (SSI) and functionally significant item (FSI) as defined in Sec. 7.5.1. Brief descriptions of the blocks in the work sheet are given below.

- *System/subsystem nomenclature:* Refers to the end product such as an aircraft or a specific subsystem.

- *Part number/FSCM/model number:* Identifier for the item. FSCM refers to Federal Supply Code for Manufacturers.

- *Mission:* The specific task the item is expected to perform.

- *Indenture level of item:* The level of the item in a functional breakdown order.

- *Zone identification:* A geographical code unique to military logistics. See App. B of Mil-Std-2080A(AS).[6]

Revision no. _____

| System/Subsystem Nomenclature | Part number/FSCM/model number | Mission | Prepared by: | Date |

Reviewed by: Date

Approved by: Date

— Item data —

| Indenture level of item | Zone identification | Number per system | Reference drawing |

| LSACN | Item description |

Compensating provisions (for each function/subfunction)

Item reliability data

MTBF (inherent) _____ MTBF (operational) _____

MTBMA _____ MTTR _____ Burn-in time _____

Minimum equipment list. (Can this aircraft or support equipment be dispatched or operated with this item inoperative? If so, list any limitations.)

Classification of item

_____ Significant

_____ Nonsignificant

Functions/ subfunctions		Functional failures (failure mode)		Failure causes (engineering failure mode)		Hidden function	Mission phase	Functional failure effects (local, next higher level, and end effects)	Functional failure detection method	Severity class	Wearout life (for each failure cause)	R & M data (for each failure cause) and Remarks
No.		Ltr.	(for each function)	No.	(for each functional failure)							

Sheet _____ of _____

Figure 7.2 FMECA work sheet for maintenance analysis.

- *Number per system:*   The number of identical items in the system.
- *Reference drawing:*   Drawing number of the item being analyzed.
- *LSACN:*   Logistic support analysis control number assigned to the item.
- *Item description:*   Description of the item in terms of its function and major components or assemblies.
- *Compensating provisions:*   Redundancies and protective features in relation to functions and functional failures.
- *Item reliability data:*   Information such as inherent and operational mean time between failures and failure rates; maintainability data such as mean time between maintenance actions (MTBMA) and mean time to repair should also be included. Other useful information: burn-in and screening requirements and safe life for limited-life items.
- *Minimum equipment list:*   Can the end item be operated if this item has failed or is inoperative for maintenance?
- *Classification of item:*   Is it significant or nonsignificant? A significant item is one whose functional failures would have safety or major economic or operational consequences. Others are nonsignificant. (Any higher-level item containing a significant item is also a significant item.)
- *Functions/subfunctions:*   List of primary as well as secondary functions, which may also have an effect on safety. For example, a primary function of a fuel pump is to pump fuel at a specified rate. A secondary function may be to contain fuel within specified boundaries.
- *Failure modes:*   A list of possible functional failures. For example, a pump at too low a flow rate may pump too much fuel, or it may not pump at all.
- *Failure causes:*   A list of all possible causes for each failure mode. For example, the causes of a fuel pump malfunction may be a drive-gear failure, a shaft failure, or a broken oil line.
- *Hidden function:*   List the functional failures that will not be evident to the operating personnel while they perform their normal duties.
- *Mission phase:*   The part of the mission in which failure would be critical. The rubber seal on a solid booster rocket may fail disastrously during lift-off, but with no effect in orbit.
- *Functional failure detection method:*   List of detection provisions that minimize the effects of failures.
- *Severity class:*   (I) Catastrophic, (II) critical, (III) marginal, (IV) minor.[7]

- *Wearout life:* List of life expectancies of the item for each failure cause.

- *R&M data and remarks:* List, based on lower-level reliability and maintainability data, of inherent and operational MTBF, MTBMA, MTTR, and burn-in/screening time (if applicable) for each failure cause. The nature of the maintenance required should be included where applicable.

7.2.6 Time-line analysis

Critical actions are usually multiple events that are either time-constrained or have safety ramifications. They require a strict protocol for their safe accomplishment. For these activities, time-line analysis provides a means to identify and analyze the sequential requirements of the maintenance and the operating tasks. The analysis determines whether the actions can be performed on a phased or parallel basis and ensures that they are in correct order of protocol. This analysis can be applied to

- Tasks that require more than one person.

- Tasks that involve safety consequences or must by nature be accomplished on a phased basis. Examples: fueling, oxygen servicing.

- Tasks that require safety equipment to be operational. Example: A flare unit that is designed to burn leaked gases from a chemical plant should not be down for maintenance during operating hours.

7.2.7 Level-of-repair analysis

Level-of-repair analysis (LORA) should be done as soon as the preliminary design is approved. The purpose of this analysis is to determine the least life-cycle costs for various alternatives involving repairing, discarding, and replacement. The analysis includes items that seem uneconomical, such as already-established safety practices and protocols.

A major portion of level-of-repair analysis deals with the level at which the repair is economical. Depending on the life-cycle costs, it may be cheaper to repair at the organization, the intermediate, or the depot level. (For life-cycle cost analysis, see Chap. 4.) Mil-Std-1390[8] contains information on level-of-repair programs.

7.2.8 Logistics support analysis documentation

In military logistics, a great quantity of records is required because of complex systems and variable manpower skills. In a commercial system, one must make a value judgment. For military systems, Mil-Std-1388-2A,[5] containing about 600 pages, spells out most of the documentation required. It suggests the following records:

Record A : Maintenance requirements

Record B : Item R&M characteristics

Record B1: Failure-modes and effects analysis

Record B2: Criticality and maintainability analysis

Record C : Operation and maintenance task summary

Record D : Operational and maintenance task analysis

Record D1: Personnel and support requirements

Record E : Support equipment and training material description and justification

Record E1: Unit under test and automatic test program

Record F : Facility description and justification

Record G : Skill evaluation and justification

Record H : Support items identification

Record H1: Application-related support items identification

Record J : Transportability engineering characteristics

Mil-Std-1388-2A, like most military standards, requires tailoring. It contains data logs so that the process can be automated. The process is iterative through all phases of the equipment life cycle and makes use of past histories.

7.3 Logistics Engineering during the Manufacturing Phase

The principles of logistics engineering apply equally well to manufacturing operations. Japan's famous kanban, or just-in-time, system would not be possible without the use of these principles. With logistics-engineered transportation and distribution, a steel plant needs to keep only 1 hour's worth of inventory (to be sure, an efficient system of quality control for materials and services must be in place). BMW in its plants in Germany introduced logistics engineering in 1976; automobile production has increased 60 percent during 1976–1986, as compared to only 20 percent for the average German manufacturer. According to Pretzch,[9] logistics at BMW is seen as a cybernetic function structured according to control-loop principles.

Computer-integrated manufacturing in many U.S. automobile plants is also a good example of logistics principles applied to manufacturing.

7.4   Logistics Engineering during the Test Phase

Without verification tests, logistics engineering problems are bound to grow by leaps and bounds. These tests can be done on mock-ups and partial systems. Sometimes verification can be done by detailed paper analysis and simulation. Generic guidelines are given below; additional tests may be indicated by analysis.

7.4.1   Tests for R&M characteristics

Reliability and maintainability characteristics tests are performed to verify MTBM, MTBF, MTTR, and other specified features. R&M-related logistics tests can be merged with the standard reliability and maintainability tests.

7.4.2   Tests of operating procedures

Operating procedure tests are conducted to see if the facilities, training, and procedures are adequate as planned. Most of these tests involve audits. All the test goals can be developed from logistic support analysis records.

7.4.3   Tests for emergency preparedness

When a major disaster strikes—a power blackout, train accident, or a major failure at an oil refinery—the maintenance crew must restore the system to operable condition under intense pressure. In such situations, technicians are likely to make mistakes, resulting in even longer downtime or another accident. For this reason, in large systems it is customary to perform emergency drills to expose deficiencies in maintainability and support requirements. The system may need a major redesign effort. For example, it often happens that rescue equipment is available, but the facility is not designed to give access to it.

7.5   Logistics Engineering in the Use Phase

Logistics engineering in the use phase involves making sure that materials, supplies, spares, training, maintenance needs and similar requirements are continually evaluated and improved. The most important function is generally maintenance. If old or degraded components are replaced, overall reliability shows little degradation with age. This section covers reliability-centered maintenance and measuring its effectiveness.

7.5.1 Reliability-centered maintenance

Reliability-centered maintenance (RCM) refers to a preventive maintenance program designed to preserve the inherent reliability of the equipment. The program emphasizes maintenance tasks selected on the basis of the reliability characteristics of the equipment and a logical analysis of the consequences of failures.

RCM analysis planning. Reliability-centered maintenance planning must be done during the design phase—not in the use phase. Since the RCM analysis is done on the basis of consequences of failure, *FMECA is absolutely necessary.* RCM may require certain features to be an integral part of the design. For example, if a noise level inside a casting is to be monitored, provisions must be made so that monitoring devices can be installed.

RCM process. The basic steps in RCM analysis are

1. Develop significant items that are critical to safety and major functions. An item may be functionally significant (loss of function would have a significant consequence on the equipment) or structurally significant (failure would result in a major reduction in residual strength of the structure). It is important to distinguish the types of significant items because different decision logic is applied to each (Figs. 7.3 and 7.4).[10]

2. Perform failure-mode, effects, and criticality analysis.

3. Analyze failure consequences.

4. Assign an appropriate maintenance task corresponding to one of the strategies below.

RCM strategies. There are at least four basic RCM strategies:

1. *On-condition monitoring:* This calls for removal or repair when condition of the part warrants it. Observation is directed at specific failure modes which give identifiable physical evidence of potential failure. (For example, if the treads on an automobile tire are less than $\frac{1}{16}$ inch thick, it may be prudent to replace the tire.) Each component is inspected at periodic intervals. It remains in service until it reaches the unacceptable condition. This strategy allows all components to re-

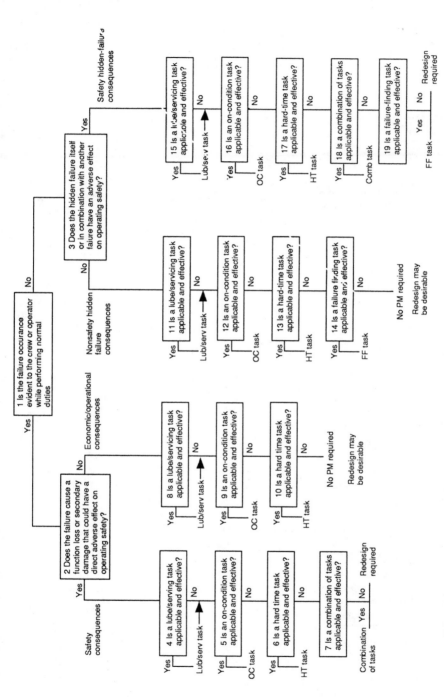

Figure 7.3 RCM decision diagram for functionally significant items.

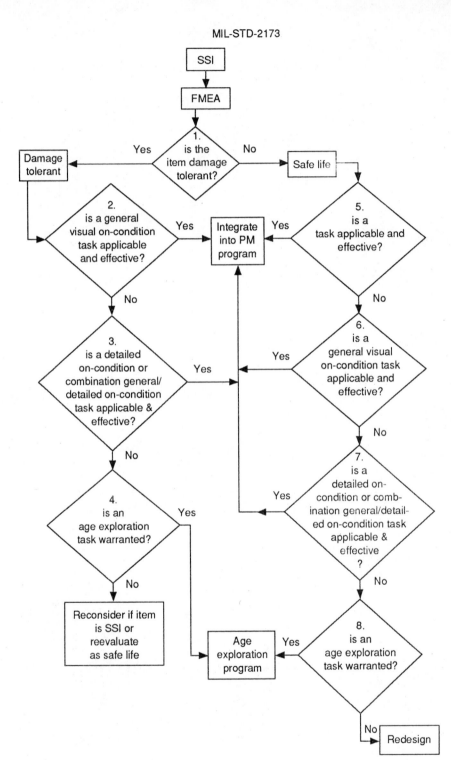

MIL-STD-2173

Figure 7.4 RCM decision diagram for structurally significant items.

alize their individual full life. Condition can also be monitored if an operator can distinguish certain sounds or smells as a warning of failure. Examples: electrical insulation may give off a smell if overheated; an automobile brake squeaks before complete wearout.

2. *Internal condition monitoring:* This strategy measures internal parameters that change before a failure. For example, electronic circuits show higher junction temperatures, which can be monitored remotely with X-ray guns or thermal scans. An electrical utility computer can monitor pressure and temperature transducers in transformers in the electrical distribution network. Acoustic emissions can be monitored to detect internal flaws in the structure of an aircraft.

3. *Hard-time tasks:* For many items, the probability of failure becomes significantly greater after a certain operating age. This is usually because of a wearout mechanism. The overall failure rate of such items can be reduced by imposing a definite ("hard") time limit on them. This prevents the use of an item at an age when the probability of failure is beyond reasonable limits.

Hard-time tasks (scheduled removals and replacements of items) are established on economic life limits when operating experience indicates that the scheduled discard or rework task is adequate on purely economic grounds. A safe-life limit for hard-time tasks is imposed only when safety is involved and no on-condition observation is practical. An item is removed *before* the specified maximum age. On aircraft structures, the safe life is usually set at one-half the minimum life, which is defined as life at which the probability of failure becomes larger than a specified value.

To determine the age-reliability relationship, a technique called *age exploration* is used. It is a process involving controlled testing and analysis of operating data. For safety-critical items, statistical analysis of probability of failure is done. In addition, the probability of unintentional events is analyzed.

4. *Group or block replacement:* This task is like a hard-time task based on the economic life of an entire group of identical items—sometimes a group of items. A simple example is the replacement of light bulbs on a busy street; a power company changed all the light bulbs during a scheduled visit rather than dispatch a maintenance crew more than 50 times a year to change one bulb at a time. Similarly, a nuclear power plant may decide to replace a block of bearings at the same time. Some companies replace several different items of roughly same age during scheduled shutdowns.

7.5.2 Measuring the effectiveness of logistics engineering

Chapter 4 describes how to assess effectiveness of maintainability engineering. A similar approach should be taken for logistics engineering. The investment in logistics engineering is productive as long as the total of all the logistics costs, when avoided costs are taken into account, is a decreasing function of time and customer satisfaction goals are met.

7.6   Topics for Student Projects and Theses

1   Write a generic design review procedure for a design for logistics support.

2   Perform a logistics support analysis for a commercial product.

3   Perform a logistics support analysis for a mass-production manufacturing process.

4   Perform a logistics support analysis for a job-shop manufacturing plant.

5   Suggest an approach for analyzing the number of spares required for a system.

6   Criticize the reliability-centered maintenance procedure. Suggest improvements.

References

1. Walker Finkelstein, "The New Concept and Emerging View of Logistics," *Logistics in Manufacturing*, John Mortimer (ed.), IFS Publications, Kempston, Bedford, U.K., 1988, pp. 201–205.
2. Eric J. Lerner, Designing for Supportability, *Aerospace America*, June 1989.
3. Charles O. Coogan, "A Systems Approach to Logistics Engineering," *Logistic Spectrum*, winter 1985, p. 22.
4. Benjamin Blanchard, *Logistics Engineering and Management*, Prentice-Hall, Englewood Cliffs, N.J., 1974.
5. *Logistic Support Analysis*, Mil-Std-1388,
6. *Maintenance Engineering, Planning, and Analysis for Aeronautical Systems, Subsystems, Equipment, and Support Equipment*, Mil-Std-2080A(AS),
7. *Procedures for Performing a Failure Mode, Effects, and Criticality Analysis*, Mil-Std-1629,
8. *Level of Repair*, Mil-Std-1390,
9. Hanns-Ulrich Pretzch, "BMW Logistics: A Step in the Future," *Logistics in Manufacturing*, John Mortimer (ed.), IFS Publications, Kempston, Bedford, U.K., 1988, pp. 145–149.
10. *Reliability Centered Maintenance Requirements for Naval Aircraft, Weapon Systems and Support Equipment*, Mil-Std-2173(AS).

Further Reading

Blanchard, Benjamin S., and E. E. Lowery, *Maintainability: Principles and Practices*, McGraw-Hill, New York, 1969.

Definitions of Effectiveness, Terms for Reliability, Maintainability, Human Factors and Safety, Mil-Std-721, Naval Publications and Forms Center, 5801 Tabor Avenue, Philadelphia, PA 19120.

DoD Requirements for a Logistic Support Analysis Record, Mil-Std-1388-2A, Naval Publications and Forms Center, 5801 Tabor Avenue, Philadelphia, PA 19120.

8

Human Factors
Engineering

8.1 Human Engineering Principles

Human errors can cause a product to fail or can result in an accident. The purpose of this chapter is to introduce the concept of *design for immunity* to human errors. A product design which is immune to errors contributes toward a *robust design*. For example, disposable cigarette lighters come in two styles: with and without a cap. A lighter without a cap is not a robust design because the flame is supposed to go out in a few seconds but sometimes does not. The user is responsible for making sure the flame is extinguished. Some people have burned their chests and faces because of such a flame. Lighters with caps do not pose this hazard. They cut off the oxygen supply, extinguishing the flame.

An investment in human factors design reduces life-cycle costs. The cost reduction is in terms of saved human lives, reduced training and skill requirements, lower liabilities, and fewer disasters.

Other examples of human errors and design for immunity are

In older automobiles, a driver can travel at speeds far above the safe limit. In many new cars, there is a limit on the top speed.

In a software project, a new programmer wrote a wrong command and caused a major disaster. The software was changed to limit the ability of the system to go beyond certain boundaries.

These examples show that mishaps can be anticipated and prevented during early design, when design changes are much cheaper. The changes are made on paper before the actual product is manufactured.

There are many products on the market which are not immune to human errors. The latest models of automobiles have radios with many small buttons and labels. In trying to read and use them, a driver may be totally distracted from the road. Indeed, several accidents have already occurred because a radio was being tuned. The operation of showers in hotels and institutions is not standardized. In some places the knob is rotated counterclockwise to increase hot water, while in other places the reverse is true. Many older people in nursing homes have received severe burns from not knowing which direction to turn the knob. But confusion would be eliminated if all designers followed the guidelines already available in Mil-Hdbk-1472.[1] Other mishaps can be prevented by performing appropriate analyses, as shown in this chapter.

8.2 Human Factors in the Design Phase

During detail design an engineer should take at least these actions (described in more detail below):

1. Use checklists and standards in developing specifications.

2. Perform design reviews for human interfaces in normal and emergency conditions.

3. Use lessons learned.

4. Review the results of hazard analyses (Chap. 5), which include hazards generated by human error.

8.2.1 Use of checklists and standards in specifications

Checklists and standards contain proven design principles. Mil-Hdbk-1472, for example, contains checklists and design guidelines that give information on display location; illumination levels; standardization of switches, knobs, buttons, and signs; workplace dimensions; access dimensions for maintenance; interpretation of warning lights; alarms and their dimensions; and much more.

This generic list of good human engineering principles can be followed:

1. Do not standardize if interchangeability will cause an accident. (A repair technician in a hospital connected an oxygen hose to a carbon monoxide source, and vice versa. The fittings were identical. They should have been different for each source.)

2. Standardize according to typical human behavior. (Usually a person turns the handle clockwise to lower a window of an automobile. But some cars are designed for counterclockwise rotation on the driver side. During an emergency, a driver might think that the window is stuck.)

3. The color blue should always be used to indicate cold and red to indicate hot conditions. (In some hotels, blue knobs have been installed for hot water. The person who installed the knobs may not have paid attention to the color code.)

4. Repetitive operations should always be automated. (They are monotonous for humans.)

5. The location of safety devices should be standardized. The emergency horn actuator in some cars is in the center of the steering wheel; in others, it is on the rim of the steering wheel; and in still others it is on a stick on the side. (A person renting a car on a business trip may not know where to push the horn in an emergency.)

6. Controls in electronic devices should be designed so that the quantity (voltage, temperature, current, and so forth) is increased in the clockwise direction.

7. The controls in hydraulic and pneumatic systems should rotate counterclockwise to increase the quantity (water, gas).

8. The color green should indicate a "go" signal, red a malfunction, yellow caution, and blinking red an emergency.

9. A design should not depend only on a lamp to give a warning, because the lamp itself may have burned out. An alternative is to keep the lamp lit all the time and have it blink in emergency. If the lamp is not lit, someone will check for the reason. Another option is to provide a backup audio alarm.

10. For remote controls, the worst situation should be considered. (Rear electric windows in automobiles have been rolled up by the drivers only to crush a child's fingers. Some new designs incorporate fiber-optic sensors for sensing fingers on the glass edge.)

11. Small, round parts, such as those in toys, can be swallowed by children. Parts should have square or rectangular shapes so that they cannot be swallowed easily.

12. Critical warnings should be large, clear, and easily noticeable. (Do not depend on the user to get this information from the operator's manual.) The warning sign should be permanently printed or secured. It should not fall off.

13. Backup devices should be clearly marked. (An operator at a nuclear plant caused an accident by trying to control an unenergized secondary pump instead of the primary pump. The control panel did not indicate which one was active.)

14. If a human can input the wrong information, such as a wrong

number on a computer panel or a keyboard, fail-safe features should be incorporated.

15. Do not design for color interpretation if any operator is color blind. Mil-Hdbk-1472[1] recommends color coding in various designs. Incidentally, most color coding involves red and green, where most color blindness exists. According to McConnell,[2] "Most color weak individuals show deficiency either in their response to red or green or to both these hues. About 5 percent of the people in the world are totally blind to one or more hues on the color circle." McConnell further points out that the main types of color blindness involve a red-green deficiency or blue-yellow deficiency. A person suffering from red-green deficiency will see the world in blue and yellow. Green grass will appear to be blue and a fire engine will appear yellow to such a person. The much smaller proportion of people with blue-yellow deficiency will see the world in red and green. Distinguishing black and white is not a problem with any type of color blindness.

16. Instruments should avoid glare and blurring on indicators. Some unfortunately allow moisture to penetrate and their dials cannot be read.

17. Knob settings should be labeled for clarity and durability. (Labels fall off, lettering fades, and glare interferes with legibility.)

18. The sequence of control use should proceed in numerical order. In one case, a system failed because the push buttons were numbered 1, 2, 3, 4, 5, 7, 6. The numbers 6 and 7 were not in sequence because of an engineering change. The operator assumed they were in sequence without reading them.

19. Positive and negative battery terminals should have different sizes to prevent incorrect cable attachment.

20. For safety-critical complex systems, the connector colors should match wire colors to prevent miswiring. (At least 14 fire control warnings on a Boeing 747 airplane were found miswired in 1989. Color coding helped prevent future problems.)

21. Numerals on instrument dials should be placed so that they are not covered by the pointer.

22. Instruments in control rooms should be mounted so that they are clearly visible to the operator. (Some controls in the Three Mile Island nuclear power plant were found behind the operator.)

23. Frequently used displays should be directly in front of the operator and controls should be on the side, within easy reach.

24. Warnings should be bright enough or should have sufficient contrast that they are legible under all illuminating conditions.

25. A computer should give positive feedback to the operator regarding acceptance or rejection of data. Otherwise, the operator may start pressing the wrong buttons in desperation.

26. An operator should not be required to remove many parts to inspect items such as filters, fan belts, and corroded parts.

27. Provision should be made to prevent accidental actuation of controls. Critical control buttons may be designed so that they do not protrude from the control panel, or guards may be provided.

28. Sharp edges should be avoided on parts to prevent cuts and lacerations.

29. Designs that require operators to be alert to dangers should be avoided.

30. Enough clearance is needed on consoles and operating equipment to avoid injuries to an operator's body parts.

31. Enough access space is needed for humans as well as tools and equipment for repairs.

32. Software that is capable of distinguishing between valid and unsafe commands should be required.

33. Humming noise in the equipment should be avoided; it may cause drowsiness or irritation.

34. If there are too many digital controls, operators cannot monitor them. Replace the critical controls with analog dials.

35. Avoid the need for verbal communications in a noisy environment. For example, when the pilot and first officer perform start-up checks in an airliner, it is difficult to distinguish if the first officer calls out "ON" or "OFF."

8.2.2   Design reviews for human interfaces

When human interface analysis (HIA) deals with a specific machine, it is called *human-machine interface analysis*. Human interface analysis may involve a repair or maintenance procedure or even a person making software program changes.

The best technique is operations and support hazard analysis. O&SHA consists of writing the operating procedure and analyzing what could go wrong with each step in the procedure. It is a technique for preventing accidents, but can be used for preventing any failure attributable to human error. For example, if the attention of an operator monitoring an instrument dial strays, an out-of-limits indication may go unnoticed. If this failure mode can cause an accident, then some other means of monitoring should be designed. An example of an O&SHA format is shown in Fig. 8.1. The procedure is the same as the maintenance engineering safety analysis in Chap. 5.

For example, suppose a failure mode is "Technician forgets to install seal on a bolt on an aircraft engine while performing preventive maintenance." The design solution is to make the seal an integral part of the bolt. The technician will never have to worry about forgetting to put the seal on.

A specialized tool called *link analysis* is used in the aircraft industry. An analyst observes which controls in the aircraft cockpit are used more frequently. Design improvements are made by placing the most frequently used controls in favorable positions.

8.2.3 Using lessons learned

A new design should not be infested with the same problems as previous ones. The information for design improvements can come from failure histories, data banks, dealers, distributors, users, and competitors. One of the best sources of information is the technique for human error rate prediction. In THERP, operators and repair personnel on similar systems are interviewed. Confidentiality is maintained; if operators do not wish to reveal their names, they are not recorded. This is a powerful technique, because most human errors are never documented but these people have experienced them and can describe them. The analyst must motivate the interviewees to talk about all real or potential problems. Many times operators may talk about the problems experienced by other operators, which is valuable information too. A good designer wants all this information. For quantifying human error rate see Ref. 3.

8.2.4 Review of hazard analyses

In Sec. 8.2.2, the operations and support hazard analysis was suggested for the human-machine interface. Actually there are several other interfaces such as the human-software interface and the human-environment interface. The most important analysis for these interfaces is preliminary hazard analysis, which covers many human-induced hazards. (Chapter 5 covers PHA and several other analyses.) A human factors specialist should be involved in coming up with any design changes resulting from the analysis. Other hazard analyses to be reviewed for human errors are maintenance engineering safety analysis (MESA) and any additional analysis with human involvement.

Gibble and Moriarty[4] take a life-cycle approach in the technique

System/subsystem title ————			Operating hazard analysis			Revision ———— Date ————	
Activity ————						Prepared by ————	
Task number	Task description	Hazard	Hazard Effect	Hazard category	Potential accident prevention measures	Proc. no.	Resolution

Figure 8.1 Example of operating and support hazard analysis.

Hazard (SSHA)	Design	Production	Maintenance	Operational
Contact C-4 shorts and causes spoilers to deploy inadvertently.	Contacts have several failure modes designed in by engineers.	Faulty processes allow defect to cause short circuit.	Maintenance worker error during installation.	Pilot overrides automatic setting.
Hazard	Design	Production	Maintenance	Operational
Ordnance fuse out-of-line; rotor allows initiation in out-of-line position.	Not enough tolerance built into rotor system by designer.	Error in production, or misassembly	Maintenance check does not detect problem.	Pilot uses wrong fuse setting.

Figure 8.2 Illustration of human factor/safety function analysis.

called *human factor/safety functional analysis*. They postulate that a human error can be initiated in almost every life-cycle phase, whether the design, production, maintenance, or operational phase. For example, Fig. 8.2 shows that short circuiting an electrical contact in an airplane is a hazard that can be introduced in any of the above phases. During design, the failure modes can be designed in by the designer. A production operator can allow a defect to cause a short circuit. Similarly, the pilot of the aircraft can override an automatic setting and cause the same accident a short circuit would have initiated.

When reviewing for human errors, the designer has to assume the worst case. No matter how careful an operator may be, there is a good chance an error will be made. The system should be designed to overcome such actions. For example, the operator may be required to use both hands to operate a machine; this assures that the operator cannot put a hand in an unsafe portion of the machine. Similarly, radio controls can be placed on the steering wheel of an automobile (Fig. 8.3). This allows the driver to use the radio, without taking eyes off the road. In addition, the steering wheel includes heater and fan controls.

8.3 Human Factors in the Manufacturing Phase

Errors by humans in manufacturing can introduce latent defects which may not be detected by inspection and test procedures. These defects affect safety and reliability of products. They may affect the safety of the humans themselves. This section covers such concerns and precautions.

8.3.1 Types of manufacturing errors and controls

Errors cited below are frequently encountered. Some suggestions for preventing them are provided.

Errors of illusion. Inspectors often see only what they want to see. If they have been seeing 100-ohm resistors on circuit boards, a 1000-ohm resistor can get by without their knowing it. If they are used to seeing a seal on an aircraft engine, a missing seal often will not be noticed. Such errors may be prevented by mechanizing inspection. In the case of a missing seal, however, mechanized inspection may not be viable. Then some kind of feedback can be designed into the system. For example, a pressure gage in the cockpit can give a warning of low pressure. Inspection may not be necessary.

Those who are not convinced about the errors the subconscious mind

Fan control   Temperature   Radio control
                       control   Radio volume
                                                                control

Figure 8.3 Radio control on a steering wheel—an enhanced human factors feature. (*Courtesy: Delco Electronics.*)

can make should try the experiment in Fig. 8.4.[5] Read the statements in the figure at your own pace and observe anything unusual or incorrect.

Over 50 percent of the people who examine the figure cannot find anything unusual in their first try! The reason: their minds are made up on what they want to see. (They do not want to be confused with facts, as the joke goes. Each of the triangles contains a redundant word—for example, "Paris in *the the* spring."

A reader who thinks this is a freak experiment should try the following.[5] Let two or more people read the phrase below and count

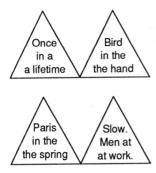

Figure 8.4   An experiment dealing with the subconscious mind. (*Source: J. M. Juran and F. Gryna, Quality Planning and Analysis, McGraw-Hill, New York, 1970.*)

the number of times the letter F appears. Very rarely will they come up with the same count.

THE FIRST FINE FISHING DAY OF THE YEAR WE FINALLY FLEW TO ALASKA FOR FIVE DAYS OF REAL FISHING.

Errors of vision.   Inspectors not only see what they expect to see; their eye retains the image of the previous piece for roughly $\frac{1}{16}$ second. This is the reason for saying that 100 percent inspection is rarely 100 percent effective. Such repetitive inspections should be done by robots.

Errors of insufficient knowledge.   This is very common. For example, a production foreman was told to change a chemical solution for cleaning steel parts before starting every shift. He was never told that the parts (for smoke detector alarms) must be very clean to prevent rust. One day, in the interest of getting higher production, he did not change the solution. The parts rusted, making the smoke detectors inoperative.

The company had to recall the alarms. Unfortunately, only 5 percent of the consumers returned the alarms. The company, according to U.S. law, was liable for the alarms still in the market. Needless to say, the company is bankrupt.

It is important therefore that the procedures clearly reflect the risks. According to assurance technology principles, the solution should have been changed by an automated device. Even better, the chemical process should have been avoided altogether by using stainless steel parts.

Errors of engineering oversight.   Production operators and inspectors are not the only ones who make mistakes. Many mistakes are attributable to oversight on the part of the engineers who design the process equipment. In one case, a soldering temperature was actually 40°F below the temperature shown on the dial. The equipment produced many cold solder joints. The oversight in this case was that the equipment designer did not give any guide for calibration. Nobody, therefore, calibrated the equipment.

8.3.2 Preventing inspection errors

Inspection is done by production operators and inspectors. In either case, certain techniques can be used. Some are

Compartmentalize the work: Divide the inspection into smaller tasks rather than one large task. If there are 300 components on a circuit board, ask the operators to check all resistors first, then the transistors, and so on. If the inspection is done according to the sequence in the parts test, it will become tedious. Another way of compartmentalizing is to divide the circuit board assembly in several sections and have all boards inspected for the first section, then for the second section, and so on until all sections are inspected.

Comparison inspection: In this method, a known good assembly is placed as a standard model and all the production is compared visually with it. If the product is complex, an automated means such as robot vision or artificial intelligence technology may be used.

Visual aids, tables, instead of paragraphs: Humans like to have information which is easily accessible. Inspection instructions several text pages long are hard to remember. Humans have proved to be much more effective when they use photographs, visual aids, and tabular columns for instructions instead of text in paragraphs.

Job sampling: Harris and Chaney[6] recommend a job sampling technique. In this technique, a known number of errors are planted (as in the error seeding technique for software) and inspection reliability is measured by the percent errors caught. This tool should not be used as a performance appraisal tool but as a training tool.

In addition to above practices, the usual motivating factors have a big role in inspection effectiveness. The supervisory environment can encourage as well as discourage errors. Sound workplace design and application of ergonomic principles will also help.

8.4 Human Factors in the Test Phase

Tests for human factors are complex because they are done systemwide. They cannot be easily done without the persons who are going to use, repair, and maintain the system. At least four tests should be done, as outlined in this section.

8.4.1 Tests for stereotype behavior

Humans act in a certain manner unless trained otherwise. The left-handed person does tasks in a certain way and so does the right-handed person. Their actions must be studied to prevent mishaps. For example, a typical way to reduce electrical load on a power system is to turn a control counterclockwise. If the control is not designed this way, then the load will increase instead of decreasing. In case of toys, as mentioned earlier, children will tend to swallow small parts. That is why a reputable toy maker makes sure the small parts are not separable from the toy or they are not round.

The best way to test for stereotype conditions is to observe the use of the system by humans on an experimental basis. A large toy maker observes children using and misusing toys through one-way mirrors. The parents are given toys and asked to report the findings. A recent test on automobiles revealed that drivers who rent cars for business purposes cannot always easily remove the ignition key from a late-model car. The car has an unusual mechanism of removal. A driver can become very frustrated if the key cannot be withdrawn; the driver should not have to struggle for minutes to figure out how to extract the key. A driver will try to remove a key in a certain natural manner; the driver should be able to remove it in that manner or be able to guess the correct procedure quickly. If 90 percent can guess and 10 percent cannot, the design is not acceptable.

8.4.2 Tests for emergency preparedness

Many disasters have been caused because, in emergencies, people have no time to think. They react to instincts and make many mistakes. One of the reasons for a blackout in New York was that a technician increased the load on the power system instead of decreasing it. In another power company, an operator was making desperate adjustments in the control room on a backup pump instead of the primary pump. Many persons in the Bhopal, India, chemical accident died because they did not know the reason for the emergency alarm. Thinking it signaled an ordinary accident, they ran toward the plant instead of away from it.

To minimize the effects of a mishap, every organization with potential for an accident should conduct emergency preparedness tests to assure that people will do what they are expected to do. These tests are called *emergency drills*. Everyone—including the people living in the vicinity—is instructed to perform certain tasks. Periodically a mock emergency is created to see how people will react in controlling the accident. Observers then interview the participants and make improvements in the system design and procedures.

Typical findings are that people cannot turn fire extinguishers on, sprinklers do not work, people turn controls in the wrong direction, handicapped people cannot escape, evacuation takes much longer than expected, and people forget to open windows in case of toxic smoke. The most frustrating finding is that no one seems to be in charge. The purpose of emergency preparedness tests is to discover all the roadblocks before a real disaster takes place.

In emergency preparedness tests, the following errors should be looked for;[7] the information will be useful for preventing future accidents.

Errors of substitution: An operator habitually selects a wrong control or device. Examples: turning on windshield wipers instead of lights in a car, turning on hot water instead of cold water in a hotel, pressing the car's accelerator instead of the brake pedal.

Errors of selection: An operator carelessly or inadvertently selects a wrong control or setting. This error is similar to the error of substitution but is not a result of habit patterns. Examples: a pilot hits a wrong switch on the cockpit controls; a computer user selects wrong commands without reading the instructions.

Errors of reading: An operator reads wrong information. Examples: an automobile driver reads the AM dial instead of FM dial on the radio; a nuclear plant inspector reads the 10-to-100 scale assuming it is the 1-to-10 scale and therefore does not detect a radiation leak.

Errors of irritation: An operator is liable to make mistakes in uncomfortable weather or when surrounded by too many confusing controls or alarms. Inability to read dials and settings also causes irritation.

Errors of warning: When caution, warning, and danger signs are not properly posted, an operator can introduce an error. Do not rely on operators to remember all the instructions.

Errors of alertness: An operator may not be alert because of lack of oxygen or presence of toxic gases, for example.

Errors of lack of understanding: An improperly trained operator may be afraid to ask questions. During emergencies this person may make too many mistakes.

Errors of haste: An operator is not able or not trained to perform work in an allocated time. In such cases wrong controls may be initiated.

Errors of sequencing: An operator may use short cuts and bypass certain operations or perform them in the wrong sequence, not knowing it may cause an accident. A DC-10 jet with more than 300 passengers crashed on takeoff at Chicago because the maintenance technicians took a short cut in mounting one of the engines.

Errors of overconfidence: It is sometimes said that a senior repair technician is more likely to be involved in an accident than a trainee. The trainee will follow the procedures while a senior technician will make independent decisions. An experienced Occupational Safety and Health Act (OSHA) inspector electrocuted himself when he checked a high-voltage power installation without wearing insulated gloves.

Errors of reversal: An operator may operate a device in the reverse direction because there is no clear indication on the control. Examples: an operator may flip a switch up instead of down, may rotate a knob counterclockwise instead of clockwise.

Errors of unintentional activation: An operator may wear oversize clothes or metal rings which may accidentally activate controls. Other examples: a driver's coat sleeve may get tangled on the speed selector and shift it to an undesired setting; long hair may get tangled in an escalator at a subway station.

Errors of oversight and omission: An operator may simply forget a task. Examples: a pilot may not call out certain checks before takeoff. A commercial airliner crashed after takeoff because a second officer did not call out for setting the wing flaps in the proper position.

Errors of mental overload: An operator may not be able to attend to critical tasks because of too many tasks, fatigue, or personal problems. Examples: too many alarms at a nuclear power plant may prevent the operator from responding to the right ones; a business traveler renting an automobile late at night may not take time to know the locations of the windshield wiper control and defroster.

Errors of physical limitations: An operator may respond incorrectly or not respond at all because of inability to reach or handle the work. There are limitations imposed by height, strength, left-handedness or right-handedness, and so forth.

Errors of casual behavior: An operator may not take the task seriously. When the driver of an automated mass-transit train left to have a chat with someone, the train took off without him. This threw passengers into panic.

8.4.3 Tests for amelioration

Tests for amelioration assess the ability to provide immediate medical and rescue services to injured people. The tests may be designed to assess

1. How long it takes to travel to a first aid location

2. How well the medical services are performed

3. How long it takes to provide emergency help to those close to death

4. How long it takes to get an ambulance

5. How long it takes to reach the hospital

6. Emergency devices available in the vicinity and in the ambulance

8.4.4 Tests for the human-machine interface

Tests of the human-machine interface can be part of emergency preparedness tests. They are done mainly to assure the effectiveness of the interface. For example, control room equipment may be tested to determine whether

- A control can be accidentally pressed and arm a weapon or activate a process at the wrong time.

- Specialized tools are secured close to the equipment so they are always accessible.

- Functions are testable individually so that the operator can identify a problem easily and quickly.

- A unit is easy to disassemble and reassemble for repair.

- Functions of knobs, switches, and lights are standardized. Sometimes two or more suppliers design different portions of a system without coordination and total confusion results; in one place the on position of a switch is up and in other places the on position is down.

- Glare on instruments hinders observation.

These and many other ideas can be obtained from standards, checklists, and personal experience.

8.5   Human Factors in the Use Phase

During normal use, many incidents happen that can result in major failures, mishaps, accidents, or near misses. They should not be taken lightly. There should be formal reporting of such incidents. Many hospitals, for example, use a data collection system based on incident/accident reports. The reports are summarized and corrective action is taken to prevent recurrence of the events.

Other errors in use are those related to maintenance and repair. Errors induced by maintenance personnel may degrade reliability and introduce unsafe conditions. Techniques such as maintenance engineering safety analysis (Chap. 5) may be used to prevent these errors.

In the use of equipment, fatigue and monotony often play an important role. In use of aircraft, human performance clearly can result in accidents.

8.6   Topics for Student Projects and Theses

1   Develop a human factors analysis approach for product reliability.

2   Develop a human factors analysis approach for quality control inspection.

3   Develop a human factors analysis approach for maintainability.

4   Develop a human factors analysis approach for manufacturing operations.

5   Develop a human factors analysis approach for software system safety.

6   Tailor a human factors analysis approach for a consumer product.

7   Tailor a human factors analysis approach for a nuclear power plant or a toxic chemical plant.

8   Develop a model for estimating human reliability for an airline maintenance technician.

9   Develop a model for estimating human reliability for a toxic chemical plant or an oil refinery.

10   Develop guidelines for including human engineering requirements in a product design specification.

11   Conduct an experiment using the job sampling technique in Ref. 6 and state your conclusions and recommendations.

12   Develop a technique for hazard analysis not covered in this chapter.

13 In Sec. 8.4.2 several types of human errors are pointed out. Can you suggest other type of errors? Develop a design checklist to avoid such errors during emergencies.

References

1. *Human Engineering Design Criteria for Military Systems, Equipments and Facilities*, Mil-Hdbk-1472, Naval Publications and Forms Center, 5801 Tabor Avenue, Philadelphia, PA 19120.
2. James V. McConnell, *Understanding Human Behavior*, Holt Rinehart Winston, New York, 1980.
3. B. S. Dhillon, *Human Reliability with Human Factors*, Pergamon Press, New York, 1986.
4. James W. Gibble and Brian H. Moriarty, "Human Factors in Accident Causation," lecture notes at University of Southern California, 1981.
5. J. M. Juran and F. M. Gryna, *Quality Planning and Analysis*, McGraw-Hill, New York, 1980.
6. D. H. Harris and F. B. Chaney, *Human Factors in Quality Assurance*, John Wiley and Sons, New York, 1969.
7. A. Bryan Leslie, Jr., "The Human Factor: Implications for Engineers and Managers," *Professional Safety*, November 1989, pp. 15–18.

Further Reading

Bass, Lewis, *Products Liability: Design and Manufacturing Defects*, Shepard's–McGraw-Hill, New York, 1986.
Hammer, Willie, *Product Safety Management and Engineering*, Prentice-Hall, Englewood Cliffs, N.J., 1980.
Johnson, W. G., *MORT Safety Assurance Systems*, Marcel-Dekker, New York, 1980.
Joint Army–Navy–Air Force Steering Committee, *Human Engineering Guide to Equipment Design*, John Wiley and Sons, New York, 1972.
Juran, J. M. *Quality Control Handbook*, McGraw-Hill, New York, 1987.
Kolb, J., and S. S. Ross, *Product Safety and Liability*, McGraw-Hill, New York, 1980.
Lupton, T., (ed.), *Proceedings of First International Conference on Human Factors in Manufacturing*, IFS Publications, Kempston, Bedford, U.K., 1984.
Sanders, M. S., and E. J. McCormick, *Human Factors in Engineering and Design*, McGraw-Hill, New York, 1987.
Swain, A., and H. Guttman, *Handbook of Human Reliability Analysis with Emphasis on Nuclear Power Plant Applications*, NUREG/CR 1278, Nuclear Regulatory Commission, Washington, D.C., 1983.

9

Software Performance Assurance

9.1 Software Performance Principles

The number of computerized medical devices recalled by the Federal Drug Administration in 1987 was double that in 1982. Among the recalled devices was a programmable pacemaker that malfunctioned when a patient walked by a store's antitheft device. Another was an x-ray machine in a hospital that produced 80 times the recommended radiation dose and killed three patients. The supplier insisted that the x-ray machine software met customer expectations because it did everything the hospital asked for in the specification. The hospital agreed that the software met the requirements, but maintained that it did not meet implied expectations. Although the hospital did not explicitly ask for a limitation on x-rays, it expected the equipment to perform in a safe manner. A court ruled in the hospital's favor and held the equipment manufacturer liable. These examples show that compliance with engineering specifications may satisfy contractual requirements but may not fulfill legal requirements. Therefore software specifications require thorough review so the engineering specifications truly reflect even the latent expectations of the user.

Assurance technologies principles apply to software just as they do to hardware, although some tools and techniques are unique to each. Tools for hardware such as FMECA, FTA, and hazard analysis can also be applied to software. Confusion about how to assure software performance comes from the lack of standard definitions of software reliability, maintainability, and other such requirements. This chapter suggests definitions that can be considered valid because of their acceptance in important contracts.[1, 2]

9.1.1 Software quality

Statements of software quality such as "The software shall be accurate and precise" or "The software shall conform to the specification" are too broad. Contracts should define software quality in terms of factors such as these:

Correctness: The extent to which a program satisfies its specifications. A general-purpose method for proving software correctness should include an easily applied and validated way of specifying assertions concerning the correct operation of the software and a way of indicating variance from correct operation (errors).

Interoperability: A measure of the ease with which one subsystem can be coupled with another. Verification procedures and testing should be as thorough for interfaces with related systems as for the main system. It is important to make sure that the users of all systems realize the implications of any interface, especially where these affect system status or timing.

Flexibility: A measure of the ease with which a program can perform (or be modified to perform) functions beyond the scope of its original requirements.

Efficiency: A measure of the use of high-performance algorithms and conservative use of resources to minimize the cost of operations.

Validity: The ability of a program to provide performance, functions, and interfaces that are sufficient for effective application in the intended user environment. The distinction between this and the definition of *correctness* should be noted. Whereas correctness pertains to specifications, validity pertains to the application as well as to the specifications.

Generality: The ability of the computer program to perform its intended functions over a wide range of usage modes and inputs, even when a range is not directly specified as a requirement.

9.1.2 Software reliability

Software reliability is the probability that software will perform its assigned function under specified conditions for a given period of time (the time here usually refers to the period during which all the user conditions are expected to occur). Reliability is not a function of degradation, since the software components do not degrade with time. This definition is a generic one. The following specific items may be included as part of reliability requirements:[2]

1. Operational requirements

- Predominantly control
- Predominantly computational
- Predominantly input/output
- Predominantly real time
- Predominantly interactive

2. Environmental requirements

   - Hardware interfaces
   - Software interfaces
   - Human interfaces
   - Degree of human interaction
   - Variability of hardware
   - Training level of operators
   - Variability of input data
   - Variability of outputs

3. Complexity considerations

   - Number of entries and exits
   - Number of control variables
   - Use of single-function modules
   - Number of modules
   - Maximum module size
   - Hierarchal control between modules
   - Logical coupling between modules
   - Data coupling between modules

4. Duty cycle

   - Constant mission usage
   - Periodic mission usage
   - Infrequent mission usage

5. Nonoperational usage

   - Training exercises
   - Periodic self-test
   - Built-in diagnostics
   - Self-maintenance

6. Qualitative characteristics

   - Correctness
   - Validity
   - Efficiency
   - Portability
   - Resilience
   - Reusability

- Fault tolerance
- Clarity
- Testability
- Readability
- Interoperability

Many of the qualitative characteristics mentioned above are defined in Sec. 9.1.1. Others are defined as follows:

Resilience: Sometimes referred to as *robustness*, resilience is the measure of a computer program's ability to perform in a reasonable manner, despite violations of usage and input conventions.

Reusability: A measure of the ease with which a computer program can be used in a different application from the one for which it was developed.

Fault tolerance: The ability of a computer program to perform correctly, despite the presence of error conditions.

Clarity: The ability of a computer program to be easily understood. Clarity is a measure not only of the computer program itself but also of its supporting documentation.

Readability: A measure of how well a skilled programmer, not the creator, can understand a program and correlate it to the original and new requirements.

Software reliability is also defined as "The frequency and criticality of program failure, where failure is an unacceptable effect or behavior under permissible operating conditions." Software reliability can be represented by the rate at which errors are uncovered and corrected. The following factors should be considered in measuring reliability:

- Does the program check for potentially undefined arithmetic operations such as division by zero?
- Are loop terminations and multiple-index parameter ranges tested before use?
- Are subscript ranges tested before use?
- Are error recovery and restart procedures tested?
- Are input data validated?
- Are test results validated?
- Does the program make use of standard library routines rather than individually developed code to perform common functions.

9.1.3 Software system safety

A simplistic definition is that a system is safe if the software will not cause the hardware to create an unsafe condition. The Tri-Service Software System Safety Working Group (Army–Navy–Air Force) defines software system safety as "The optimization of system safety in the design, development, use, and maintenance of software and its integration with safety-critical systems in an operational environment." Associated definitions are

Safety-critical computer software components or units: Those components or units whose errors could result in a hazard.

Safety-critical functions: Functions in hardware, software, or both whose errors could result in a hazard.

Safety-critical path: A path through the software leading to the generation of a safety-critical function.

9.1.4 Software maintainability

Software maintainability is the probability that a program will be restored to working condition in a given period of time when it is being changed, modified, or enhanced.

Software maintainability may be defined qualitatively as the ease with which software can be understood, corrected, adapted, tested, and enhanced. Quantitatively, maintainability can be assessed indirectly by measuring the following maintainability attributes: problem recognition time, administrative delay time, maintenance-tool collection time, active correction (or modification) time, local testing time, global testing time, maintenance review time, and total recovery time. It is a catchall term used to summarize all the features of a program that allow it to be easily altered or expanded. It includes these factors:

Modifiability: A measure that takes into account the extent to which likely candidates for change are isolated from the rest of the computer program. (Example: Isolation of input and output routines that are hardware- or human-dependent would increase a program's modifiability.)

Portability: A characteristic of software that allows it to be used in a computer environment different from the one for which it was originally designed. Use of standard high-level languages is one of the ways to increase portability. It can also be defined as the ability of computer software to run with different kinds of hardware and software. Portability has become an important aspect of maintainability. Newer and superior hardware is being introduced so rapidly that companies are willing to pay extra for this feature.

Availability: The percentage of a specified time period that a system is in an operable state. It can be calculated by dividing the system's uptime by total scheduled time. It can be reported weekly, monthly, or annually. Although availability is sometimes used as a measure of maintainability exclusively, actually it is a measure of reliability and maintainability together.

Testability: A characteristic of a program that allows its functional requirements to be logically separated for step-by-step testing.

Modularity: A measure of the number of independent parts in a software system. A high modularity helps to localize design modifications.

9.1.5   Software logistics engineering

Although the term *software logistics engineering* has seldom appeared in the literature, it is an important concept nonetheless. It may be defined as that branch of engineering that deals with software support functions such as maintenance, trainability, preparation of technical manuals, security, and preservation of software.

9.1.6   Some important definitions

Since the software field is still groping for standardization in many areas, it will be helpful to define some terms not already covered. They are compiled largely from Refs. 3, 4, and 5.

Coding: The translation of a detailed design representation to the appropriate programming language.

Computer hardware: Devices capable of accepting and storing computer data, executing a sequence of operations on computer data, or producing any output data (including control outputs).

Computer software: A combination of associated computer instructions and computer data definitions that enables the computer hardware to perform computational or control functions.

Computer software component (CSC): A distinct part of a computer software configuration item (CSCI). CSCs may be further decomposed into other CSCs and computer software units.

Computer software configuration item (CSCI): Software that is designated by the procuring agency for configuration management.

Computer software unit (CSU): An element specified in the design of a computer software component that is separately testable.

Coupling: A measure of the strength of module interconnection; modules with strong interconnections are highly coupled and those with weak interconnections are loosely coupled.

Deliverable: Documentation associated with each step in the software engineering process.

Design walk-through: An informal design evaluation by an independent reviewer, usually another programmer familiar with the project.

Do-while: A structured control flow construct that repeats a sequence of statements while a specific initial condition is true. Note that the sequence of statements is not executed if the specific condition is initially false.

Firmware: Software that resides in a nonvolatile medium that is read-only in nature, and is completely write-protected when functioning in its operational environment.

Hazard: An inherent characteristic of a thing or situation which has the potential of causing a mishap. Whenever a hazard is present, the possibility, regardless of degree, of a mishap occurring exists.

Hazardous operation/condition: An operation (activity) or condition (state) which introduces a hazard to an existing situation without adequate control of that hazard, or removes or reduces the effectiveness of existing controls over existing hazards, thereby increasing mishap probability or potential mishap severity, or both.

Higher-order language: Englishlike user-oriented languages oriented toward problem solving rather than detailing the work of the machine.

IF-THEN-ELSE: A control structure that implements conditional instruction execution.

Independent verification and validation (IV&V): An independent test and evaluation process that ensures the computer program satisfactorily performs, in the mission environment, the functions for which it was designed. Verification is the iterative process of ensuring that, during each phase of development, the software satisfies and implements only those requirements that were approved at the end of the previous phase. Validation is the test and evaluation process to ensure that the software meets all system and software performance requirements.

Integration test: The process of testing several unit-tested modules as a whole, to assure compliance with the design specification.

Lines of code: The actual number of instructions at the source code level of a programming language that collectively perform a software function.

Maintenance: The last phase of the software life cycle that begins when software is operational and released to one or more users; includes the supervision of the software configuration as well as the physical modification of the software.

Mishap: An unexpected, unplanned, or undesired event or series of events which has harmful consequences. Harmful consequences are injury; death; occupational illness; damage to or destruction of equipment, facilities, or property; or pollution of the environment.

Morphology: Shape of a software structure; measured in terms of depth, width, fan-out, and fan-in.

Operating system: Software that controls the execution of computer programs and that may provide scheduling, debugging, input-output control, accounting compilation, storage assignment, data management, and related services.

Preliminary design review (PDR): A formal, technical, and management review of the software development effort that concentrates on the top-level structural design and its traceability to the requirements.

Product baseline: Configuration items which have design frozen at established program milestones (system design review, preliminary design review, critical design review), and are ultimately subjected to formal testing and configuration audits prior to delivery.

Program design language (PDL): A language that describes a design using a combination of structured programming constructs and English prose.

Requirements review: A formal review of the software requirements specification to determine whether the software requirements specification (SRS) is acceptable to both developer and requester and adheres to the system specification and the software plan.

Safety-critical computer software component (SCCSC): Computer software component (unit) whose inadvertent response to stimuli, failure to respond when required, or response out-of-sequence or in unplanned combination with others can result in a critical or catastrophic mishap, as defined in Mil-Std-882B.

Software: All instructions, logic, and data, regardless of the medium on which they are stored, that are processed or produced by automatic machines and which are used to control or program those machines. Software also includes firmware and documentation associated with all of the above.

Software configuration: An instantaneous snapshot of the physical representation of software at some point in time. This representation takes two forms: (1) nonexecutable material generated to document or complement the computer program and (2) executable material processed directly by a computer.

Software requirements specification (SRS): A document that concentrates on four aspects of a software project: information flow and interfaces, functional requirements, design requirements and constraints, and testing criteria to establish quality assurance.

Stub: A dummy procedure used to test a superordinate procedure.

Support software: All software used to aid the development, testing, and support of applications, systems, test, and maintenance software.

System specification (SS): The document that defines overall system requirements without a detailed regard for the implementation approach. The document specifies functional characteristics and performance objectives of the system, interface characteristics, environment, overall design concepts, reliability criteria, design constraints, and predefined subsystems.

9.2 Software Performance in the Design Phase

Studies show that about 60 percent of software errors are logic and design errors; the remainder are coding- and service-related errors. With such a high magnitude of errors from the design phase, at least 40 percent of the software budget should be allocated to the design phase. Many service-related errors also can be prevented by proactive designs.

9.2.1 Software quality assurance in design

Since the emphasis is on accuracy and precision, the entire software development process must be monitored. The most widely used procedure is identified in DOD-Std-2167A[3] (Fig. 9.1). It is intended to in-

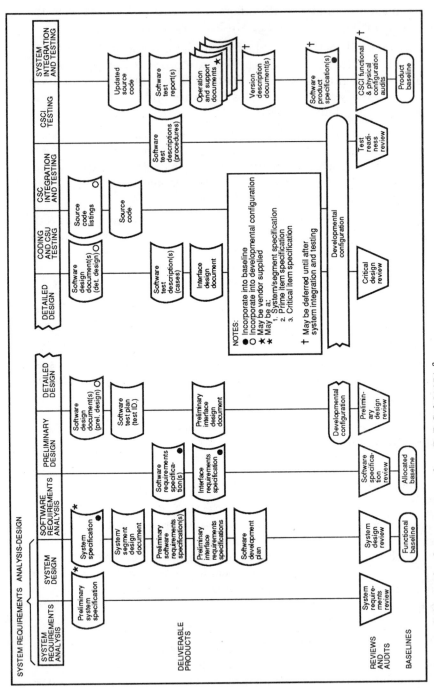

Figure 9.1 Software development process from DoD-Std-2167A.[3]

clude reliability and safety features, but these items are not detailed. The procedure shows that the quality assurance function starts in the system definition phase. At this stage, the system performance— that of hardware and software together—is analyzed. Then the software system specification is developed along with an interface specification. At this time, the functional baseline is established. Design reviews are held at each stage of the development and a baseline established after each major development, all the way to acceptance testing.

9.2.2 Software reliability in design

While errors in hardware often can be seen, software errors are invisible except to the programmer. Tiny bugs can have disastrous effects. They have killed sailors, devastated space programs, and threatened to collapse stock markets. These bugs breed fast in complex software, where thousands of lines of code are trusted to hundreds of programmers who work only on a module or two and have limited information on interfacing modules. The vast majority of systems are deeply flawed from the viewpoint of reliability.

Software has no random failures and has infinite life, once it works. There is no wearout. The traditional hardware testing strategies therefore may not work. Often the right test combinations do not occur until much later during use. Most software engineers agree that practically 100 percent of software defects are human errors, mainly in three areas: specification, coding, and modification. We have depended on the intuition of the programmer, who does not necessarily perform design reviews, and logic designers, who do not perform fault-tree analyses. "Such language-specific smart editors promise at most freedom from syntactic errors and simple semantic errors," according to Fred Brooks, Kenan Professor of Computer Science at the University of North Carolina.[6]

In view of the complexity of defining a failure, one has to consider two kinds of definitions. One is explicit and is in the contract. The second definition is implied. In safety-related incidents the courts tend to accept the hidden implied definition. The following definitions of *failure* and *fault*[7] may be considered explicit.

A failure is

1. The termination of the ability of a functional unit to perform its required function.

2. The inability of a system or system component to perform a required function within specified limits. A failure may be produced when a fault is encountered.

3. A departure of program operation from program requirements.

4. The event, or inoperable state, in which any item does not, or would not, perform as previously specified.

A fault is

1. An accidental condition that causes a functional unit to fail to perform its required function.

2. A manifestation of an error in software. A fault, if encountered, may cause a failure. It is synonymous with *bug*.

3. An immediate cause of failure (for example, maladjustment).

4. A condition, action, or element of software that, when encountered during test or operation of the system or components, results in a failure.

Software design techniques for reliability.   Derating, safety margins, and fault tolerance are design provisions for hardware reliability. Somewhat similar provisions exist for software reliability. A study reported in Ref. 2 suggests these approaches:

1. *Control complexity:*   This is the most important factor to be controlled in software. Higher complexity tends to make computer programs incomprehensible. It affects other factors such as readability and modularity. Software must be reviewed, discussed, and tested by programmers to assure the complex relations can be understood by the user. Complex software is prone to errors and makes it difficult to avoid faults. On the other hand, it allows many functions to be performed with less memory per function. Usually the reliability per function is increased. Such tradeoffs must be carefully evaluated.

2. *Perform top-down functional decomposition:*   This structural breakdown of computer software configuration items helps keep the code traceable and readable. The generic hierarchy is system, subsystem, program, module, routine, unit, and instruction. The following hierarchy applies to government contracts:

    a. Computer software configuration item (software aggregate for configuration control).

    b. Computer software component (functionally or logically distinct part of a CSCI).

    c. Unit (lowest-level logical entity which completely describes a nondivisible function to allow independent implementation and testing).

    d. Module (lowest physical entity which may be assembled or com-

piled alone). *Note*: If the module is a logical entity, it will appear higher in the hierarchy as a *unit*.

  e. Instruction (single executable statement of code for a single action).
  f. Operation (action to be performed by the instruction).
  g. Operand (addresses in the computer memory where data reside).

3. *Modularize:*   The segmentation of computer programs into single-purpose, single-entry, single-exit modules is the most popular technique discussed in the literature.

4. *Design hierarchically:*   Most authors acknowledge the significant reduction in complexity that can be achieved by enforcing a hierarchical structure of module segment calling and controlling relationships.

5. *Follow structured approaches:*   The advantages of structured techniques are well known. They tend to simplify complex software. Top-down or chain structures are examples.

Preventing specification and design errors.   Software teams can use design reviews. Software fault-tree analysis (SFTA) is very productive, but its use is not widespread. The industry has traditionally relied on so-called IV&V—independent validation and verification. A tool known as *soft-tree analysis*, which applies to hardware/software interfaces, has been tried successfully. But the problem is, as Patrick Hoff of the Naval Surface Weapons Center puts it, "These tools can require intimate workings of the software. It is at this point that many or even most engineers throw in the towel. In spite of multidisciplinary engineering backgrounds, most are not prepared for this type of analysis."

This author used failure-mode, effects, and criticality analysis with the help of programmers and influenced many changes in specifications. FMECA, in fact, is very powerful if used on functional modules. Boeing has developed software for software sneak circuit analysis.[8] All these analytical tools and many more often reveal weaknesses in specifications, particularly with regard to testability, fault masking, fault tolerance, clock synchronization, and maintainability. Section 9.2.6 contains an example of one such tool (software system failure-mode and effects analysis) applied to software reliability.

9.2.3   Software maintainability in design

Many programmers have learned through hard experience that, unlike hardware, software often deteriorates in reliability when a failure is fixed. This is especially true for software having more than a few thousand lines of code. The reason is that a programmer repairing a module often is not aware of all the interface effects and ends up creating more serious errors. Good configuration control may provide some relief, but writing good documentation and using interface traceability matrices are even more effective.

The major tools for maintainability are modularity, top-down structured programming, and sequential programming as opposed to spaghetti coding. In top-down structure the branches of the software are not allowed to be connected sideways to other branches. To search for an error, one has to look below only for a fault. Ensuring modularity and testability, and using fault-isolation techniques are also powerful aids to maintainability.

Artificial intelligence tools can help too. For example, a tool being developed at the Navy Materiel Command, Dover, N.J., called *software fingerprinting*, sets limits to complexity in the modules and establishes a profile ("fingerprint") of the baseline. The program identifies critical paths and attempts to limit their number to less than 10. Since no two profiles are supposed to be alike, each profile becomes a fingerprint. The set of fingerprint parameters can consist of source code, object code, module function, program flow mapping, and similar items. Changes in any of these items will result in a fingerprint change. If an unexpected change takes place in any module, its fingerprint becomes different, and it can be identified instantly. Among many other artificial intelligence techniques is the use of expert systems that can detect multiple failure modes and determine the effect of changes on the system.

Those who are concerned about controlling software errors must allocate time and financial resources to analytical tasks. It is cheaper to prevent the problem than to spend 100 times more during troubleshooting. The following ranking done by Chenoweth and Schulmeyer[9] on a Rome Air Development Center (RADC) study gives some insight into sources of errors. The data show the importance of interface design. The user-interface and routine-to-routine interface errors add up to 13.32 percent!

Source of error	Portion, %
Logic	21.29
Input/output	14.74
Data handling	14.49
Computational	8.34
Preset database	7.83
Documentation	6.25
User interface	7.70
Routine-to-routine interface	5.62

Even though software maintainability is a qualitative attribute, General Electric[5] defines some quantitative measures:

1. Problem recognition time
2. Administrative delay time
3. Problem diagnostic time
4. Maintenance tool collection time
5. Specification change time
6. Active correction time
7. Local testing time
8. Regression/global testing time
9. Change review time

9.2.4 Software system safety in design

The definition of system safety (Sec. 9.1.3) implies that the prevention of software hazards must be done during the entire life cycle, with software and hardware analyzed jointly.

In order to proceed with hazard analysis, a *software hazard* needs to be defined. By itself the software does not cause any harm. On the other hand, by itself it is of no use. It only becomes useful—as well as dangerous—when it is integrated with hardware. Therefore, a software hazard is one that can introduce error or malfunction resulting in an unsafe condition in the hardware.

The U.S. Air Force Inspection and Safety Center[10] classifies software hazards into four broad categories:

1. An inadvertent or unauthorized event which can lead to an unexpected or unwanted event.
2. An out-of-sequence event; a planned event occurs but not when desired.
3. Failure of a planned event to occur.
4. The magnitude or direction of event is wrong. This is normally caused by an algorithm error.

Some possible hazards in software are

- Routines which disable interrupts.
- Conditions that call for an exact match of floating-point numbers (this can result in an infinite loop).

- Conditions that result in division by zero. Many times a very small number may be truncated to zero by the software.

- Precision and scaling of data resulting in hazardous condition.

Software safety risk assessment. In hardware, risk is defined as a product of severity and frequency for each potential accident. In software, once the bug is out, it is not expected to appear again. Therefore, risk may be mainly assessed in terms of severity ratings. The Tri-Service Software System Safety Working Group has recommended the following categories:

I. Software exercises autonomous control over potentially hazardous hardware systems, subsystems, or components without the possibility of intervention to preclude the occurrence of a hazard. Failure of software or failure to prevent an event leads directly to an occurrence.

IIa. Software exercises control over potentially hazardous hardware systems, subsystems, or components, allowing time for intervention by independent safety systems to mitigate the hazard. However, the systems by themselves are not considered adequate.

IIb. Software item displays information requiring immediate operator action to mitigate a hazard. Software failures will allow or prevent the hazard's occurrence.

IIIa. Software item issues commands over potentially hazardous hardware systems, subsystems, or components requiring human action to complete the control function. There are several redundant, independent safety measures for each hazardous event.

IIIb. Software generates information of a safety-critical nature that is used to make safety-critical decisions. There are several redundant, independent safety measures for each hazardous event.

IV. Software does not control safety-critical hardware systems, subsystems, or components and does not provide safety-critical information.

Software safety tools. The U.S. Air Force *Software System Safety Handbook*[10] suggests the following safety analysis tools:

1. *Preliminary and follow-on software hazard analysis:* This is done at the concept stage. It is a brainstorming technique to identify portions of the software such as modules or routines that can create hazards. Analysis is done on system specifications, subsystem specifications, interface specifications, functional flow diagrams, lessons-

learned data, program structures, and information related to testing, coding, storage, modification, and use.

2. *Software fault-tree (soft-tree) analysis:* In this technique the hardware and software trees are constructed with standard symbols (Chap. 5). The two trees are then linked at their interfaces so that the entire system can be analyzed. A slightly different way of accomplishing this is to construct a hardware failure tree in which the software interfaces are identified. A software fault tree is then added to the interfacing points.

3. *Use of cut sets:* This technique identifies single-point and multiple-cause failures and is shown for hardware in Chap. 5. In case of the software system the technique is applied to soft trees. Computer programs for cut set analysis are commercially available.

4. *Common cause analysis:* Examination of cut sets can reveal some causes which can cause the system to fail in spite of many redundancies. A cut set is a path or a set of components which, if it fails, can cut off the system function. These are also referred to as *common causes* for the downstream failures. In the functional block diagram in Fig. 9.2, cut set 1 is the common cause of failure of components B through H to operate. Similarly, the failure of component D is the common cause of failure of redundant components E, F, G, and H to operate. Chapter 5 shows a cut set algorithm based on fault trees. For example, if a power supply is common to two data registers, a failure

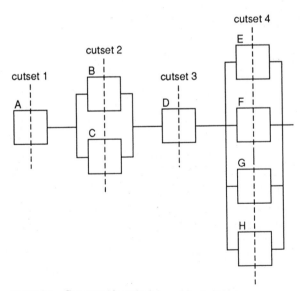

Figure 9.2 Cut sets identified for a block diagram.

in it will disable both data registers. Similarly, a voltage transient in the power line can cause a malfunction in several portions of a system. Computer programs are available for this analysis.

5. *Software sneak circuit analysis:* See description of this technique in Chap. 5 for hardware.

6. *Use of Petri nets:* This is a mathematical model invented by Carl Petri in 1961. One of the advantages is that the user can describe the model graphically. The system is modeled in terms of conditions and events, and traces the conditions as events take place in time. In a way it is a simulation technique and requires complex software. Readers interested in this technique may refer to work of Nancy Leveson.[11]

7. *Use of Mil-Std-882B[4]:* By far the most widely used guidelines are in this standard. It requires analysis and design improvements in several stages. Hardware hazard analysis procedures are covered in Chap. 5 and are equally applicable to software.

The software tasks recommended in Mil-Std-882B are

Task 301: Software requirements hazard analysis: This task is done at the very early design stage. Its purpose is to examine system and software requirements and design to identify unsafe modes such as out-of-sequence events, wrong events, inappropriate magnitudes, inadvertent commands, deadlocking, and failure to command. Documents such as system specifications, interface specifications, and flow charts must be analyzed and design changes recommended.

Task 302: Top-level design-hazard analysis: The results of task 301 are compared in analyzing the top-level design. Some further analysis is accomplished. Safety-critical software components are identified and their degree of risk is estimated. The test plans are developed. This analysis should be complete and approved before detailed design is started. At this time, the results of various hardware hazard analyses may be integrated to identify issues that require more analysis, questionable elements in the specification, and interface concerns that may lead to unsafe conditions.

Task 303: Detailed design hazard analysis: The above analysis is repeated at detailed design level to identify software-critical computer software components that must be brought under configuration control as required in DoD-Std-2167A. The analysis is done at lower levels of details such as databases, files, tables, and variables. Also developed at this stage are operator, diagnostic, user, and detailed test requirements. These analyses are then passed on to code developers.

Task 304: Code-level software hazard analysis: This analysis begins when code writing begins. It continues throughout the life cycle. Reviews of safety-critical codes are made in detail and code changes are evaluated for consistency with the results of previous hazard analyses. Among many improvements, the following may be considered: fail-safe requirements, input overload, out-of-bound conditions, error default handling, recovery procedures, hardware failures, independent reviews, handling of transients, error detection, error corrections, fault tolerance, and fault marking.

Task 305: Software safety testing: Software must be tested sufficiently to assure the hazards identified during hazard analysis are no longer present or at least are controlled to an acceptable level. Software is tested within the specified boundaries as well as under abnormal conditions.

Task 306: Software/user interface analysis: This task assures that the user will have clear information and that a mistake made by the user will not result in a software hazard. After this task, the design will include giving clear indication of hazard to the user, provision for recovery from the hazard, safe cancellation of an event, and safe override features.

Task 307: Software change hazard analysis: The purpose of this analysis is to assure safe maintenance of software. All changes to specifications, requirements, design, code, systems, equipment, test plans, descriptions, procedures, cases, and criteria are subjected to hazard analysis and testing unless a reason for it to be unnecessary is known.

One can notice that software analysis tools are still vague. "How to" guides practically do not exist. Captain Steven Mattern[12] conducted a survey of the users of software safety analysis tools (the users are few), with the help of the Tri-Service Software System Safety Working Group. The results are given in Table 9.1.

Software fault-tree analysis. The listing of hazard analysis tools in Table 9.1 shows that fault-tree analysis ranks highest in number of users. The consensus is that for most effective use, fault-tree analysis should be applied to both hardware and software as components of one system. Figure 9.3 shows a fault tree for a hardware-software-human system. All three elements are analyzed in the same tree.

For a safe and fault-tolerant design, the criterion is that the failure of the system should not be a result of a single-point failure. There-

TABLE 9.1   Survey Results

Software hazard analysis tool	Users
Fault-tree analysis	8
Software preliminary hazard analysis	4
Traceability matrices	3
Failure modes and effects analysis	3
Requirements modeling/analysis	2
Source code analysis	2
Test coverage analysis	2
Cross-reference tools	2
Code and module walk-throughs	2
Sneak circuit analysis	2
Emulation	2
Subsystem hazard analysis	2
Failure-mode analysis	1
Prototyping	1
Design and code inspections	1
Checklist of common software errors	1
Data flow techniques	1
Hierarchy tool	1
Compare and certification tool	1
System cross-check matrices	1
Top-down review of code	1
Software matrices	1
Thread analysis	1
Petri net analysis	1
Software hazard list	1
BIT/FIT plan	1
Nuclear safety cross-check analysis	1
Mathematical proofs	1
Software fault-hazard analysis	1
Mil-Std 882B series 300 tasks	1
Topological network trees	1
Critical function flows	1
Blackmagic	1

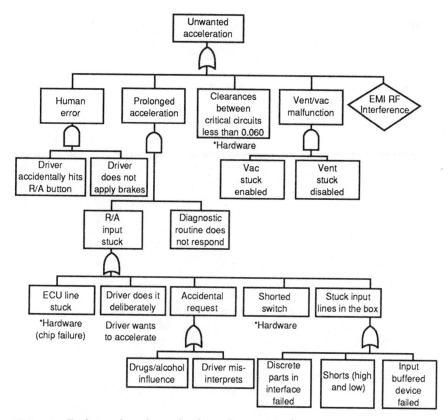

Figure 9.3   Fault tree for software-hardware-human interface.

fore, the top level should always have an AND gate. In the system of Fig. 9.3, AND gates were designed into the product at the second level because the AND gate at the top was not viable. When this criterion is used, many faults at lower level can be tolerated. For one design, about 4 months of analysis was eliminated at the lower levels because of the protection. Designs using this criterion have been safe and efficient and result in lower life-cycle costs.

9.2.5 Software logistics engineering

Support provisions, including trained programmers and software engineers to change, modify, or enhance the software, are part of software logistics engineering. So, too, is documentation. User manuals should be started at the same time as coding. They should not be written at the last moment and must be tested as part of system tests.

Another aspect of logistics engineering is to plan and execute data collection. A closed-loop feedback system should be used during the

entire software development process to assure high reliability and
maintainability. Data analysis should continue after the software is
operating. It will provide valuable information for the next project.

9.2.6 Software system failure-mode and effects analysis

Software system failure-mode and effects analysis (SSFMEA) applies
to software systems where software controls the hardware. Most technical products belong in this category, such as aircraft, power plants,
refineries, and complex systems. In SSFMEA, the focus is on identifying system weaknesses through software flow charts so that software
specifications can be made complete, clear, comprehensive, and unam-
biguous. The goal is

1. To determine whether the software is fault-tolerant with respect to
   hardware failures

2. To identify missing requirements in the system specification

Usually hardware improvements are also required for the software
to be effective. A software engineer should never assume that the
hardware will work perfectly; it does not. Many times there are
tradeoffs to be made between hardware and software improvements.
SSFMEA emphasizes software, although there is no harm done if
hardware is improved as a by-product. The reason for emphasizing
software is that software errors result in enormous downtime and the
causes are not obvious to the user. Hardware errors, in contrast, are
often so obvious that even nontechnical people can easily notice them;
if a tire on an automobile is stolen or flat, anyone can detect it. Of
course, if the hardware seems to be unusually unreliable, one can focus on hardware flow charts. Ideally, one team focuses on software
while an independent validation and verification team focuses on
hardware.

The objective. The main objective of SSFMEA is to enhance the role of
software in assuring the safety and reliability of the system. In doing
so, the hardware system is also enhanced because the software will
not be able to do its job without the help of fault-tolerant hardware.
The second objective is to reduce software life-cycle costs. SSFMEA
analysis is done as soon as the specification is written and is continually updated as the design work progresses, well before coding. This
allows many engineering changes to be done in a short time and at the
lowest cost.

The methodology. SSFMEA is simple. A functional flow chart from
the software specification is developed. Each block in the flow chart is

treated as a component of the software system. Then the potential hazards in the system are analyzed during execution of each software component. An attempt is made to control the hazard and critical failures through software changes or hardware control, whichever offers a viable solution and lower life-cycle costs.

Demonstration of methodology. Consider a simple example of SSFMEA. The flow chart in Fig. 9.4 shows four software components in an aircraft. For the *check engine speed* component, the system failure modes and causes are

1. Software receives erroneous speed reading. *Causes*: Wrong bit stuck in the microprocessor; sensor inaccurate.

2. Software receives no inputs on speed. *Causes*: Sensor failed; microprocessor failed.

3. Software is OK; all hardware is OK, but no report appears on the printer. *Causes*: Instrument response cycle may be longer than the software waiting time; printer malfunction.

Once the causes of faults and errors are identified, the design improvements become more obvious. Consider failure mode 3, where everything is all right except that the software does not report the status. The potential cause is that the instrument response time may be longer and software may move on without reacting to it. The *effect* of

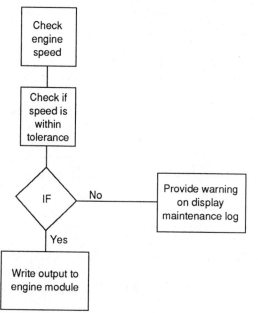

Figure 9.4 Four components of a software item.

this functional failure is that the maintenance log will receive a warning of an out-of-tolerance condition which is a false alarm. The software design in such a case should be improved if the criticality of this failure is high. The following design precedence should be applied:

- Remove cause of the hazard from the design
- Design for fail-safe consequences
- Design for fault tolerance
- Provide self-checking software every time a critical operation is performed
- Implement special training

In this example, the cause of the hazard can be removed by having the software wait until the instrument responds. One may argue that the software waiting time was based on the instrument response time in the first place, so why should we worry? We still need to worry because the response time can change as the instrument gets older. Sometimes additional circuits are connected to the instrument which can change the response time. The upshot is that, if the instrument does not respond in a reasonable time, then the failure of the instrument should be reported—instead of a false alarm.

Figure 9.5 shows a partial flow chart of an aircraft early warning system, and Fig. 9.6 shows an SSFMEA format for the software components. As noted, the SSFMEA will identify missing requirements, as is evident in this example. The first software element in Fig. 9.6, *Check exhaust temperature*, is associated with the failure mode *Erroneous input* because of the malfunction of the sensor. The current design requires that the deviation should be reported on the printer. From an SSFMEA point of view, reporting a hazard may not be good enough; the jet engine may fail in midair before it receives maintenance. The proper action is to switch over to a redundant sensor. This is the missing requirement that should be added to the system specification. Similarly, the third failure mode, *Computational error* caused by voltage transients, should be *forgiven* by the software design. A common technique in such cases is the use of *time redundancy* which calls for running the same algorithm at three different times; at least two of three outputs must match. Thus, if an error is temporary, it will not match the remaining two outputs.

Software-hardware interface control.   Often hardware failure will prevent software from functioning. A typical software engineer's response might be, "Well, if the hardware has failed, it is not a software problem." This argument is not accepted in the courts. The legal system

CONDITION MONITORING

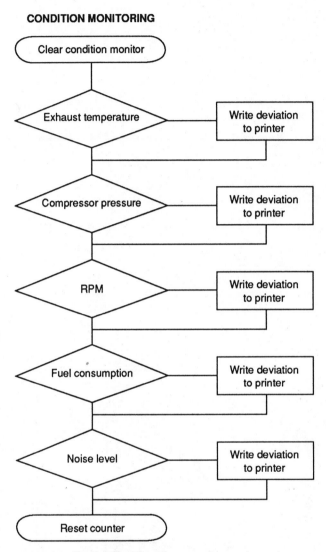

Figure 9.5 Partial flow chart for an early-warning condition monitoring system in a jet engine.

does not differentiate between hardware and software. It is concerned with the system. It is the responsibility of the manufacturer to make the system perform in spite of the hardware failure. Consider software controlling a fan through a microprocessor to cool an engine. If the microprocessor has failed, the software is helpless. The question is can the software still control the fan? The answer is yes, if the software

Software element	System failure mode	Cause	Effects	Criticality	Recommended software/hardware changes
1. Check exhaust temperature (within tolerance)	a) Erroneous input	Sensor malfunction	software reports large deviation	IIC	a) Software should switch to redundant sensor, when large deviation is reported. Add redundant sensor.
	b) No input	Sensor failed	same as (a)	IIC	b) Same as (a)
	c) Computational error	Voltage transient Temporary fault	same as (a)	IIB	c) Use time redundancy
	d) Temperature reading very high	Exhaust temperature	engine fails	IB	d) Software to alert pilot to shut down engine and report on maintenance log.
2. Check compressor pressure	a) Pressure too low	Clogged or dirty filter, microprocessor error	possible engine failure	IC	a) Software to alert pilot to shut down engine and report on maintenance log.
	b) Pressure too high	Compressor malfunction	possible engine failure	IC	b) Same.
	c) Compressor OK but erroneous pressure reported	Algorithm error, sensor malfunction	false alarm	IIA	c) Include independent check on compressor pressure.

Figure 9.6   SSFMEA format for the early-warning system in Fig. 9.5.

and hardware engineers want to work together. There are many ways the software can be kept in charge. One way is to provide an external switch through which the software can turn on the fan.

Other uses of SSFMEA. So far we have discussed only SSFMEA's role in design productivity. However, SSFMEA can be used for many other purposes and should be documented in detail in these applications.

One important use is in developing test cases. Many companies spend thousands of dollars in developing test cases and are still unable to do a thorough job because it takes a monumental amount of time and money to check all paths. SSFMEA points out important failure modes, and at least the test cases related to these failures ought to be developed. Other tools such as soft trees can identify additional test cases.

SSFMEA can be used for software maintainability and availability analysis. Since it predicts failures, the associated downtimes can be estimated. If the downtimes are not acceptable, the software can be redesigned to include fault tolerance, modularity, and early diagnostic features.

Implementing an SSFMEA program. Implementing hardware FMEA has taken years in some cases and has still been done wrong (This is discussed in Chap. 3). Because it has been practiced wrong for many years, people assume that the wrong way is the right way. Most often, people violate the first rule of FMEA, that the analysis be performed by a team; in almost every defense contract, FMEA is done by one person who is really not familiar with the internal working of the design, and therefore not qualified. But in a team, the same individual can be very effective by being curious and asking a lot of questions based on personal experience. An effective SSFMEA is one performed by a team of hardware, software, and system engineers with guidance from a system safety or reliability engineer.

A major misuse of FMEA has been that it is used mainly for predicting failure modes and reliability. These are not the primary goals of FMEA. The real goal—to make design improvements—is mostly overlooked. Some perform FMEA so late that no one is allowed to make major engineering changes. The investment in FMEA for these situations is not justified. The only way to make it dramatically effective is to do it very early in design. And since only management can make it happen, the process must be institutionalized.

9.2.7 Software performance specification

The features discussed so far should be an integral part of a software specification. Unfortunately, companies rarely spend enough time on developing a clear and complete specification, and, as a result, many designers and programmers spend more time on testing and fixing the problems. That is why the software is often well behind schedule. Figure 9.7 presents guidelines used by a large company for writing performance specifications. Additional items can be added as appropriate.

9.2.8   Software design review checklist

In any design review, a checklist is a good place to start. Every company should develop its own checklists for software quality, reliability, safety, and maintainability. This author found a checklist (Fig. 9.8) from a major aerospace company to be one of the best. It was developed mainly for software system safety but can be applied to reliability. The list is slightly modified to include some additional features.

9.3   Software Requirements during Coding and Integration

During the coding and integration stage, plans are made to prevent coding errors. One of the best ways to reduce errors is to standardize the programming language, but there has been only limited acceptance of a standard language. The Pentagon standardized on the Ada language for all weapon systems so that reliable software components can be put together, just as in hardware systems. The effort has been successful, but it limits the ability of military systems to work with commercial software.

9.3.1   Coding errors

Many types of errors are rampant in codes. A programmer can remove many errors by searching for them by type. Major error categories are

1. *Computational errors:*   Result from coded equations, algorithms, and models.

2. *Configuration errors:*   Code is incompatible with the operating system or the application software. Generally occurs after a modification of the software.

3. *Data handling errors:*   Made in reading, writing, moving, storing, and modifying data.

4. *Documentation errors:*   Mistakes in software descriptions, user instructions, and other documents.

5. *Input/output errors:*   Result from input and output code.

SYSTEM SPECIFICATION

INTRODUCTION

Software is often an integral part of a larger configuration that includes hardware, facilities, and personnel. This configuration, called a "system," must be carefully specified prior to commencement of software planning and development.

The SYSTEM SPECIFICATION indicates the overall functional, operational, and performance characteristics of a system. In it both hardware and software elements are described and the requirements for each are specified.

The software organization is normally not responsible for the generation of the SYSTEM SPECIFICATION. However, because the system document serves as a baseline for all software documents, software personnel should participate in its development.

A recommended format for the SYSTEM SPECIFICATION is presented in this Appendix.

1.0 SCOPE

This section provides an overview of the system and defines scope for performance, design, development and test requirements. In addition, system objectives and characteristics are discussed.

2.0 APPLICABLE DOCUMENTS

This section lists all documents which apply (in whole or in part) to the specification.

3.0 REQUIREMENTS

This section contains a descriptive and quantitative definition of all system requirements. The following topics are considered:

1. The performance and design requirements for the system.

2. The requirements related to operating, maintaining, and supporting the system, to the extent these requirements define or constrain design of the system.

3. The design constraints and standards necessary to assure compatibility of system items.

4. The definition of the principal interfaces between the system being specified and other systems with which it must be compatible.

5. The functions of the system, and the principal interfaces between and within each function.

6. The allocation of functional performance and the definition of specific design constraints.

7. The identification and use of existing equipment, computer programs or operating procedures.

Unless purely descriptive by nature, requirements are stated in quantitative physical terms with tolerances that can be verified by subsequent tests, demonstrations, or inspection. Requirements stated in this section are the basis for the tests specified in Section 4 of the specification.

3.1 System Definition

This paragraph lists the elements to be developed or included as part of the system. Elements include hardware, software, and operating procedures. A graphical representation (block diagram) of these items should also, be shown if appropriate.

Figure 9.7 Guidelines for writing system specifications. (*Source: General Electric Co., Software Engineering Handbook, McGraw-Hill, New York, 1986.*)

3.1.1 Functional Description

This paragraph contains a brief description of the system including a list of all functions. The intended use of the system is described to the extent necessary for an overview understanding.

3.1.2 System Diagrams

This paragraph incorporates the system level functional schematic diagrams and includes the top-level functional control and data flow of the system. The level of detail that is required must identify all system functions and their relationship with system elements.

3.1.3 Interface Definition

This paragraph describes the functional and physical interface: 1) between the system and other systems with which it must be compatible; and 2) between all functions within the system.

3.1.4 Existing Resources and Procedures

This paragraph lists existing elements which the system is to be designed to incorporate. This list identifies the elements (e.g., hardware, software) by reference to nomenclature, specification number and/or part or model number. If the list is extensive, it may be included in an appendix which can be referenced in this paragraph.

3.1.5 Operation and Organization

This paragraph includes an overview description of the intended operation of the system. Also, the anticipated deployment of the system both geographically and organizationally (e.g., the number and location of installations) is described.

3.2 Characteristics

System characteristics are described in detail in the paragraphs that follow.

3.2.1 Performance Characteristics

This paragraph contains system performance characteristics and provides sufficient guidance for technical development. It describes what the system should accomplish and specifies both upper and lower performance limits. The following considerations are included:

1. Quantitative criteria covering endurance capabilities of the equipment that are required to meet the user needs under stipulated environmental and other conditions. Minimum total life expectancy is stated.

2. Quantitative expected performance for each operational mode (e.g., average rates, peak rates, quantity of inputs).

3. Programming language requirements (for software), if language is dictated by other system elements.

4. Other essential aspects of system performance which cannot be more appropriately located under another heading.

3.2.2 Physical Characteristics

This paragraph considers the following topics (as required):

1. Weight limits, electrical and equipment heat generation specifications.

2. Dimensional limitations, space, operator station layout, and access for maintenance.

3. Requirements for transport and storage, such as tie down, pallets, packaging, and containers.

4. Durability factors to indicate degree of ruggedness.

5. Health and safety criteria, including consideration of adverse explosive, mechanical, and biological effects.

6. Security criteria.

7. Media format for delivered software.

3.2.3 Reliability

Reliability is stated in quantitative terms. The conditions under which the reliability requirements must be met are explicitly defined.

3.2.4 Maintainability

This paragraph specifies maintainability requirements. The requirements apply to maintenance in a planned maintenance and support environment and are stated in quantitative terms. Examples are:

1. Time (e.g., mean and maximum downtime, mean time between failures, reaction time, turnaround time, mean and maximum time to repair, mean time between maintenance).

2. Rate (e.g., maintenance man-hours per specific maintenance action, maintenance hours per operating hour, frequency of preventive maintenance).

3. Maintenance complexity (e.g., number of people and skill levels, variety of support equipment).

Fig. 9.7   (*Continued*)

3.2.5 Availability

The degree to which the system is available, e.g., 24 hours per day, 7 days per week.

3.2.6 Environmental Conditions

This paragraph includes environmental conditions to be encountered during operation and/or storage of the system equipment. The following subjects should be considered: natural environment (wind, rain, temperature, etc.); induced environment (motion, shock, noise, etc.); electromagnetic signal environment.

3.2.7 Transportability

This paragraph includes requirements for transportability which are common to all system equipment to permit installation and logistic support. All system equipment that will be unsuitable (due to operational or functional characteristics) for normal transportation methods is identified.

3.3 Allocation of System Functions

This paragraph allocates the system functions described above to hardware, software or other system elements. Minimum system design and construction standards are also specified.

3.3.1 Division of Hardware and Software

This paragraph contains a list of all system functions and the allocation of either hardware or software.

3.3.2 Materials, Processes and Parts

This paragraph specifies requirements that govern the use of materials, parts, and processes. It contains specifications for particular materials and processes to be utilized in the design of system equipment. In addition, requirements for the use of standard components and parts for which a qualified products list has been established are specified.

3.3.3 Workmanship

This paragraph contains workmanship requirements for equipment to be produced during system development. Requirements for manufacture by specified production techniques are also discussed.

3.3.4 Interchangeability

This paragraph specifies the requirements for system elements that are interchangeable and replaceable. Entries in this paragraph are for the purpose of establishing a condition of design, and are not to define the conditions of interchangeability.

3.3.5 Safety

This paragraph specifies safety requirements that are basic to the design of the system. Equipment characteristics, methods of operation, and environmental influences are considered.

3.3.6 Human Engineering

Human engineering requirements for the system are specified and applicable documents (e.g., regulations) included by reference. Particular attention is paid to areas in which the effects of human error would be particularly serious.

3.4 Logistics

This paragraph considers the logistical characteristics that are required for the implementation, support and maintenance of the system.

3.4.1 Maintenance

This paragraph includes consideration of factors such as: 1) use of test equipment; 2) repair versus replacement criteria; 3) organizational levels of maintenance; 4) maintenance and repair cycles; 5) accessibility, and 6) distribution and location of spare parts.

3.4.2 Support

This paragraph specifies the support required for the system. Considerations include vendor hardware and software support and maintenance response requirements.

3.4.3 Facilities and Equipment

This paragraph specifies the impact of the system on existing facilities and facility equipment. It also describes requirements for new facilities, auxiliary equipment or software to support the system. Software development facilities and test facilities are also included.

3.5 Personnel, Training and Documentation

This paragraph specifies the training requirements for the system and includes: 1) how training is to be accomplished (e.g., school, on the job); 2) identification of equipment or software that will be required for training purposes; and 3) course material and training aids. Additionally, a training schedule and location are specified.

3.5.3 Documentation

This paragraph specifies the documentation required for each system element. References to standards are acceptable.

Figure 9.7 *(Continued)*

4.0 QUALITY ASSURANCE

Provisions and criteria established in this section are used to verify functional and performance requirements specified in Section 3. Requirements for test/validation are defined. The following topics should be considered:

1. Collection and recording of data during all testing for use as part of the reliability analysis;

2. engineering evaluation and test requirements in direct support of design and development activity;

3. integration testing such as continuity checking, software/hardware interface mating, and function operation in the installed environment; support equipment compatibility and documentation verification;

4. formal test and validation of functional and performance characteristics.

5.0 NOTES

This section contains background information and other material pertinent to a complete description of the system.

6.0 APPENDIX

This section contains supplementary material that is part of the specification but has been appended for convenience in specification maintenance.

Figure 9.7 *(Continued)*

Item	Req'd	Not Req'd	Incorporated
General and Miscellaneous			
Provides for precluding dependence on administrative procedures			
Provides for using information control concept for deriving the authorization code for the activation of the authorization device			
Provides that the software contains only features or capabilities required by the system, and that it does not contain additional capabilities, e.g., testing, troubleshooting, etc.			
Provides for positive control of system safety-critical functions at all times			
Provides for safety-critical subroutines and subprograms to include "Come From" checks to verify that they are being called from a valid calling program			
Separation of Commands, Functions, Files, and Ports			
Provides for using separate launch authorization and separate launch control functions to initiate a missile launch			
Precludes the ground ordnance enabling arming code from being same as the enable launch authorization code			
Provides for requiring separate "arm" and "fire" commands for ordnance initiation			

Figure 9.8 Example of a software system safety checklist.

'Item	Req'd	Not Req'd	Incorporated
Separation of Commands, Functions, Files, and Ports			
Precludes using input/output ports for both critical and noncritical functions			
Provides for sufficient difference in addresses for critical input/output ports versus noncritical ports that a single address bit failure does not allow access to critical functions or ports			
Provides for having files that are unique and have a single purpose			
Interrupts			
Provides for defining specific interrupt priorities and responses			
Provides for software system management of interrupt control so as to not compromise safety-critical operations			
Shutdown, Recovery, and Safing			
Provides for a fail-safe recovery from inadvertent instruction jumps			
Shutdown provisions are included in software upon detection of unsafe conditions			
Provides for the system reverting to a known predictable safe state upon detection of an anomaly			
Provides for software safing of safety-critical hardware items			
Provides for an orderly system shutdown as the result of a command shutdown, power interruptions, or other failures			
Requires that the software be capable of discriminating between valid and invalid external interrupts and shall recover to a safe state in the event of an erroneous external interrupt			
Provides for entry into a safe state in the event of erroneous entry into a critical routine			
Protects against out-of-sequence transmission of safety-critical function messages by detecting any deviation from the normal sequence of transmission. When this condition is detected, the software terminates all transmissions, recycles to a known safe state, and displays the existing status so the operator can take compensatory action			

Figure 9.8 (*Continued*)

Item	Req'd	Not Req'd	Incorporated
Shutdown, Recovery, and Safing			
Provides for initializing all unused memory locations to a pattern, that if executed as an instruction, will cause the system to revert to a known safe state			
Provides for identifying safing scenarios for safety-critical hardware and including them into the decision logic			
Provides for the capability of reversing or terminating launch authorization and ordnance arming functions			
Preventing, Precluding, and Disallowing Actions			
Provides for preventing inadvertent generation of critical commands			
Provides for disallowing coexistence of potentially hazardous routines			
Provides for preventing bypass of safety devices during test			
Following computer memory loading, automatic control is prevented until all data is loaded and verified			
Precludes inadvertent operation of data entry control to critical routines			
Provides for precluding a change in state if data synchronization is lost			
Provides for prevention of a hardware failure or power interruption from causing a memory change			
Provides for prevention of memory alteration or degradation over time during use			
Provides for program protection against unauthorized changes			
Provides for not allowing the safety-critical time limits in decision logic to be changed by the console operator			
Provides for preventing inadvertent entry into a critical routine			
Provides for not allowing a hazardous sequence to be initiated by a single keyboard entry			
Prohibits transmission of any critical command found to be in error and notifies the operator of the error			

Figure 9.8   (Continued)

Item	Req'd	Not Req'd	Incor-porated
Shutdown, Recovery, and Safing			
Provides that controlling or monitoring of nuclear weapons be incapable of bypassing operator control of safety-critical functions			
Provides for disallowing use of workaround procedures when reverting to a safe configuration after the detection of an anomaly			
Provides for not using a "stop" or "halt" instruction or causing a CPU "wait" state. The CPU is always executing, whether idling with nothing to do or actively processing.			
Provides for detection and termination of commands requesting actions beyond the performance capability of the system			
Provides for disallowing performance of a potentially hazardous routine concurrently with a maintenance action			
Memory, Storage, and Data Transfer			
Precludes storage, in usable form, of information required to cause initiation of a safety-critical function			
Provides for self-test capability to assure memory integrity			
Provides for prevention of a hardware failure or power interruption from causing a memory change			
Provides for prevention of memory alteration or degradation over time during use			
Provides for erasure or obliteration of clear text secure codes from memory			
Provides for limiting control access to storage devices memory			
Provides for protecting the accessibility of memory regions dedicated to critical functions			
Provides for having safety-critical operational software instructions resident only in nonvolatile read-only memory			
Provides for not using scratch files for storing or transferring safety-critical information between computers			
Provides that remote transfer of data cannot be accomplished until verification of data to be transferred is accomplished and authorization to transfer the data has been provided by the operator(s)			

Figure 9.8   (*Continued*)

Item	Req'd	Not Req'd	Incorporated
Memory, Storage, and Data Transfer			
Provides for self-test capability to assure memory integrity			
Verification and Validation Checks			
When a test specifies for the removal of safety interlocks, the software provides for verification of reinstatement of these safety interlocks at the completion of the testing			
Provides for verification and validation of status flags			
Requires that critical data communicated from one CPU to another be verified prior to operational use			
Provides for software validation of critical commands			
Provides for verification of the existence of prerequisite conditions prior to command issuance in accordance with predefined operational requirements			
Provides for verification of the results of safety-critical algorithms prior to use			
Provides for verification of safety-critical parameters or variables before an output is allowed			
Decisioning verifies the sequence and logic of all safety-critical command messages and rejects commands when sequence or logic is incorrect			
Provides that remote transfer of data cannot be accomplished until verification of data to be transferred is accomplished and authorization to transfer the data has been provided by the operator(s)			
Provides that all operator actions that set up safety-critical signals are verified by software based on control device positions			
Provides for control of analog functions having feedback mechanisms that provide positive indications of the function having occurred			
Provides for verification and validation of the prompt for the initialization of a hazardous operation or sequence of hazardous operations			
Provides for verification of accomplishment of each step of a hazardous operation, or sequence of hazardous operations, by setting of a dedicated status flag prior to proceeding to and initiating the next step in the operation or series of operations			
Provides for verification/validation of all critical commands prior to transmission			

Figure 9.8 (*Continued*)

Item	Req'd	Not Req'd	Incor-porated
Logic, Structure, Unique Codes, and Interlocks			
Provides for identification of flags to be unique and single purpose			
Provides for using unique arming codes to control critical safety devices			
Provides for inclusion of system interlocks			
Provides for using a minimum of two separate independent commands to initiate a safety-critical function			
Provides for the majority of safety-critical decisions and algorithms to be contained within a single (or few) software development module(s)			
Provides for single CPU control of a process which can result in major system loss, system damage, or loss of human life to be incapable of satisfying all of the requirements for initiation of the process			
Requires that decision logic using registers which obtain values from end-item hardware and software not be based on values of all "ones" or all "zeroes"			
Requires that decision logic using registers which obtain values from end-item hardware and software use specific binary patterns to reduce the likelihood of malfunctioning end-item hardware/software satisfying the decision logic			
Provides for cooperative processing between launch control point and missile computer(s) to process safety-critical functions			
Provides for having safety-critical modules with only one entry and one exit point			
Provides for having files that are unique and have a single purpose			
Provides for not having operational program loads contain unused executable code			
Monitoring and Detection			
Provides for inclusion of monitoring of safety devices			
Provides for detection of inadvertent computer character outputs			
Provides for detection of errors during computer memory loading to terminal loading process			
Provides for detection of unauthorized operation of data entry control			

Figure 9.8   (*Continued*)

Item	Req'd	Not Req'd	Incor- porated
Monitoring and Detection			
Provides for identification of safety-critical functions requiring continuous monitoring			
Provides for detection of improper processing that could degrade safety			
Provides for detection of a fault having the potential of degrading safety			
Provides for detecting a predefined safety-critical anomaly and informing the operator what action was taken			
Requires that the software be capable of discriminating between valid and invalid external interrupts and shall recover to a safe state in the event of an erroneous external interrupt			
Provides for detection of improper sequence requests by the operator			
Provides for detection of inadvertent transfer of safety-critical routines			
Provides for detection and termination of commands requesting actions beyond the performance capability of the system			
Reasonableness Checks			
Provides for software system reasonableness checks of all safety-critical inputs			
Provides for performing parity on other checks, requiring two decisions, before providing an output			
Initialization, Timing, Sequencing, and Status Checking			
Provides for a status check of critical system elements prior to executing a potentially hazardous sequence			
Provides the proper configuration of inhibits, interlocks, safing logic, and exception limits at initialization			
Provides for issuance of good guidance signal subsequent to satisfaction of performance of flight safety checks			
Provides for timing sufficiency of commands relative to response to detected unsafe conditions			
Provides for software initialization to a known safe state			
Provides for performing a status check of safety-critical elements prior to executing a potentially hazardous sequence			

Figure 9.8 (*Continued*)

Item	Req'd	Not Req'd	Incor- porated
Initialization, Timing, Sequencing, and Status Checking			
Provides that all critical timing relative to hazardous operations processing is automated			
Provides for employing time limits for operations impacting system safety and having these time limits included in decision logic			
Protects against out-of-sequence transmission of safety-critical function messages by detecting any deviation from the normal sequence of transmission. When this condition is detected, the software terminates all transmissions, recycles to a known safe state, and displays the existing status so the operator can take compensatory action			
Provides for initializing all unused memory locations to a pattern, that if executed as an instruction, will cause the system to revert to a known safe state			
Provides for matching of observed flight terrain being matched to computer-stored flight terrain map prior to issuance of critical commands (e.g., arming, fire, climb, descend, etc.)			
Applies use of software timing coincident with hardware timing to prevent initiation of safety-critical functions			
Provides for verification and validation of the prompt for the initialization of a hazardous operation or sequence of hazardous operations			
Provides for verification of accomplishment of each step of a hazardous operation, or sequence of hazardous operations, by setting of a dedicated status flag prior to proceeding to and initiating the next step in the operation or series of operations			
Operator Responses and Limitations			
Requires an operator response for initiation of any potentially hazardous sequence			
Provides for not allowing the safety-critical time limits in decision logic to be changed by the console operator			
Provides for concise definition of operator interactions with the software			
Provides for operator cancellation of current processing in a safe manner			
Requires that an operator cancellation of current processing be verified by an additional operator response			

Figure 9.8 (*Continued*)

Item	Req'd	Not Req'd	Incor-porated
Operator Responses and Limitations			
Provides that controlling or monitoring of nuclear weapons be incapable of bypassing operator control of safety-critical functions			
Requires that the system responds to predefined safety-critical anomalous conditions by notifying the operator of the condition and identifying the action taken			
Provides that upon safing the system, the resulting system configuration or status be provided to the operator and await definition of subsequent software activity			
Provides that remote transfer of data cannot be accomplished until verification of data to be transferred is accomplished and authorization to transfer the data has been provided by the operator(s)			
Provides that operator control of safety-critical functions is maintained under all circumstances throughout weapon system operation			
Provides that all manual actions that set up safety-critical signals are verified by software based on control device positions			
Operator Notification			
Requires that an override of a safety interlock be identified to the test conductor by a display on the test conductor's panel			
Provides for generation of critical status to operator			
Provides to operator identification of overrides to safety interlocks			
Provides for software indication if unauthorized action has taken place			
Provides for the system informing the operator of the anomaly detected			
Provides system configuration status to operator upon safing of safety-critical hardware items			
Provides for positive reporting of changes of safety-critical states, e.g., absence of an armed indication does not constitute a safe condition			
Provides for detecting a predefined safety-critical anomaly and informing the operator what action was taken			
Provides for the software system to display safety-critical timing data to the operator			

Figure 9.8 (Continued)

Item	Req'd	Not Req'd	Incor-porated
Operator Notification			
Provides for the software systems to indicate to the operator the currently active operation(s) and function(s)			
Provides for identification to the operator that a safing function execution has occurred; provides the reason for the execution with a description of the safing action taken			
Provides for notification of improper keyboard entries by the operator			
Prohibits transmission of any critical command found to be in error and notifies the operator of the error			
Provides that upon safing the system, the resulting system configuration or status be provided to the operator and await definition of subsequent software activity			

Figure 9.8 (*Continued*)

6. *Interface errors:* Occur in (*a*) routine-to-routine interface, (*b*) routine-to-system software interface, (*c*) file processing, (*d*) user interface, and (*e*) data base interface.

7. *Logic errors:* Design errors.

8. *Operating system errors:* Occur in operating system software, the compiler, the assembler, and the special utility software.

9. *Operator errors:* Human-induced errors.

10. *Database errors:* Residual errors in the database, usually from the user.

11. *Recurrent errors:* Duplicates of errors, usually arising when problems are reopened.

12. *Requirements compliance errors:* Software fails to provide capability specified in the requirements specification.

13. *Definition errors:* Improper definition of variables or constants.

14. *Intermittent errors:* Caused by temporary faults in the hardware or software.

9.3.2 Quantifying software errors

Demand for higher software reliability has created the need for measurement models. For hardware, data banks can offer some insight into reliability prediction, but software reliability is heavily process- and human-dependent. Little has been published on predicting software reliability from historical data. However, at least 67 mathematical models for software reliability have been published. Almost all are variations of two curves (Fig. 9.9).[13]

The experience of Sagols and Albin[14] suggests that the first model applies to simpler software. The second model is for complex software where the learning process is longer. Some software data do not fit either of these models. In such cases, the most common approach is to use the data on number of errors at time t to predict the number of errors remaining. Some compute mean time between failures based on

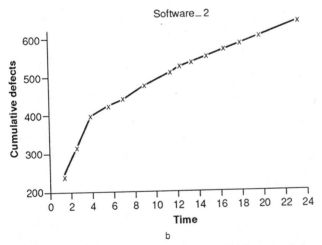

Figure 9-9 Two basic models of software reliability. (*a*) S-shaped curve (*b*) Standard exponential growth with negative coefficient.

the exponential model. Reference 2 classifies some models as below. Readers interested in more details on models should consult Ref. 14.

Reliability measurement models
- Hecht measurement model
- Nelson model

Reliability estimation models
- Reliability growth model (La Padjla)
- Mills model
- Rudner model
- Musa model
- Jelinski-Moranda deeutrophication model
- Jelinski-Moranda geometric deeutrophication model
- Jelinski-Moranda geometric Poisson model
- Schick-Wolverton model
- Shooman exponential model
- Weibull (Wagoner) model
- Goel-Okumoto bayesian model
- Littlewood-Verrali bayesian model
- Shooman-Natrajan model
- Shooman-Trivedy Markov model
- Shooman micromodel
- Littlewood Markov model
- Littlewood semi-Markov model
- Moranda a priori model
- Hecht estimation model

Reliability prediction models
- Motley and Brooks model
- McCall, Richards, and Walters model
- Halstead model

9.3.3 Coding error prevention

There are many approaches to reducing coding errors: high-level languages, such as assembly language and Ada; redundant algorithms; microcoding for complex algorithms; multiuser support for editing, compiling, and unit testing; compiler conformance to American National Standards Institute (ANSI) document X3.9-1978; user support in establishing detailed requirements, testing, and acceptance; error detection codes; independent verification and validation; system integration testing; independent testing; use of a software support library; configuration control; thorough software development plans; rigidly controlled requirement specifications; rigidly controlled software functional specifications; requirements traceability matrixes; rigidly controlled detailed design specifications; structured analysis tools; program specification language; program design language; automated test tools; computer-aided software engineering; peer walk-throughs; progress reviews; quality audits; software problem reports; specification change notices; software requirements reviews; test readiness reviews; extensive interface reviews; and human interface requirements.

Fortunately there are several software development tools available that can detect features of a software language. (An example is Utility in the UNIX operating system, which detects bugs in C language.) The tools perform much more thorough checks than a compiler can. Some detect bugs in the program files. They identify unreachable statements, loops not entered at the top, variables declared and not used, and inconsistencies in function calls. With the current trend to portability of software to new platforms, such tools have started playing a very important role in high-quality software development.

A good rule of thumb for avoiding many coding errors is "Those who design software should not write the code; those who write the code should not do the testing."

Example 9.1 Some coding errors can be prevented by using a simple integration certification form (Fig. 9.10). This makes sure the integration is done properly. The important coding errors should be caught in the process.

Example 9.2 To assure that codes will not cause any major mishaps, the review team should identify the undesirable paths that should not occur and the important paths that must occur. These paths are entered in a matrix along with test cases to test them. Such a matrix is shown in Fig. 9.11.

9.4   Software Testing

Tests are conducted for quality, reliability, safety, and maintainability. Tests may also be done for overall qualification.

9.4.1   Testing for quality

Tests for quality are aimed at measuring the accuracy and precision with which the software performs. Acceptance is based on the software's ability to perform according to specifications.

Some common quality control tests are defined below:

1. *Black box testing:*   A testing strategy that derives test data solely from the requirements specification.

2. *White box testing:*   This is a complement of black box testing: a

Processor(s): _____DP_____

Unit name: ___GP-PART-UPDATE___   Release/version: __1.0___

Test certified by: _____   Date: _____

Testing: Yes   No

1. Are all necessary elements present?   ___   ___

    a. Software design notes
    b. Test cases/scripts
    c. Test drivers?

2. Do the test cases adequately test the program   ___   ___
    logic or intergration dataflow?

3. Is the data used sufficiently close to real-   ___   ___
    world data?

Comments:

Figure 9.10   Example of a certification form for test integration.

strategy that derives test data from knowledge of the program's internal structure.

3. *Gray box testing:*   This strategy derives test data by combining elements of black box and white box testing.

4. *Unit testing:*   Individual testing of each module by its implementing programmer to verify its correctness.

5. *Validation testing:*   A formal test and evaluation of the software to demonstrate its conformance to the requirements specification.

6. *Boundary value testing:*   A testing strategy that derives test data from boundary conditions, i.e., situations directly on, just above, and just below the boundaries of input and output equivalence classes.

9.4.2 Testing for reliability

Since reliability tests are of very long duration, mathematical models are often used as part of either software field assessment or accep-

Unit Name __GP-PART-UPDATE__   Task ___DP_____   Date _____

			Test Certifications →											

			Test Case Coverage															
Path	Path 10	Path Condition	1	2	3	4	5	6	7	8	9	10	11	12	13	14	15	16
1	DF-T	Do for "NUM-XFER" entries	X	X														
2	IF-T	Found HDR entry in XFER table	X	X														
3	IF-T	Found FILE-ID, FILE-TYPE match	X	X														
4	IF-F	Match not found																X
5	IF-T	Found TRLR entry in XFER table	X	X														
6	IF-T	Found FILE-ID, FILE-TYPE match	X	X														
7	IF-F	Match not found																X
8	IF-F	No HDR or TRLR entry																X
9	DF-F	All entries complete	X	X														
10	IF-T	Error from GP-DIR-UPD																X
11	IF-F	No error from GP-DIR-UPD	X	X														
12	IF-T	FILE-TYPE "UPD"	X	X														
13	IF-F	Invalid FILE-TYPE																X
14	IF-T	Error in FMBPEN																X
15	IF-F	No error in FMPPEN	X	X														
16	IF-T	Error in FMP position																X
17	IF-F	No error in FMP position	X	X														
18	DF-T	Do for "NOM-XFER" entries	X	X														
19	IF-T	Found data entry in XFER-TAB	X	X														
20	IF-T	Found FILE-ID, FILE-TYPE match	X	X														
21	DF-T	Do for all words in record	X	X														
22	DF-F	All words complete	X	X														
23	IF-T	Error from FMPWRITE																X
24	IF-F	No error from FMPWRITE	X	X														
25	IF-T	Error from FMPCLOSE																X
26	IF-F	No error from FMPCLOSE	X	X														
27	IF-F	No match of FILE-ID, FILE-TYPE																X
28	IF-F	No data entry in XFER-TABLE																X
29	DF-F	All num-XFER entries complete	X	X														
30	IF-T	Error from GP-DIR-UPD																X
31	IF-F	No error from GP-DIR-UPD	X	X														
32	IF-T	TABLE-ID exists	X															
33	IF-F	No TABLE-ID		X														
34	DF-T	Do for delete "NUM-XFER" entries	X	X														
41	DF-F	All entries deleted	X	X														

Figure 9.11   Example of a test matrix for desired and undesired paths.

tance testing. In many military contracts, Mil-Std-781,[15] which is based on an exponential probability distribution, is also used for acceptance testing for hardware as well as the software. In some contracts, if the cause of error is not isolated in hardware or software, it counts as two errors.

Testing can be accomplished in many ways. Individual modules may be tested with stubs as the software develops and integrated later. At the system level, testing may be done bottom-up or top-down or may start in the middle (sandwich testing). Myers[16] suggests several such tests for quality and reliability.

9.4.3 Testing for maintainability

Most maintainability parameters are measurable, such as modularity, testability, flexibility, and transportability. The tests have to be developed from the customer requirements, since very little is published in this area.

9.4.4 Testing for software safety

Special safety tests are done to verify the performance of safety-related test descriptions, procedures, and cases. They include verification of inhibits, interlocks, and traps. Tests are conducted under specified and abnormal conditions. The tests are documented with special care because they can be summoned in the courts in case of injuries or deaths.

Special consideration should be given to testing of interfaces: the hardware-software interface, software-software interface, and software-human interface. The inputs to safety tests come from various hazard analysis techniques covered in this chapter.

Safety tests should go beyond normal testing to allow for safety margins. The following should be developed:[17]

- Tests to verify that the software responds correctly and safely to credible single and multiple failures (and faults)
- Tests for operator input errors and component failures
- Tests to assure that errors in data received from external sensors or other software processes will be handled safely
- Tests for failures associated with entry into, and execution of, safety-critical software components
- Tests for negative and no-go testing
- Tests to assure that the software performs only those functions for which it was intended, and no extraneous functions

9.4.5 Testing for overall qualification

Qualification testing is generally done in stages. It starts with unit testing and goes all the way to operational testing. Howley[18] suggests the following tests. Not all tests are required for every project; the selection will depend on the test objectives (e.g., validating logic) and the design or operational performance capabilities.

1. *Checkout:*   Involves compiling or assembling and executing the program, ascertaining performance characteristics, and performing error detection.

2. *Unit/module testing:*   Includes verifying design requirements and performing error detection.

3. *Integration testing:*   Verifies the compatibility of composite program elements.

4. *Computer program testing:*   Consists of verifying the functional characteristics of the program, demonstrating the validity of processing results, and establishing the validity of specifications. This level of testing includes verification and formal validation. Acceptance and quality assurance testing may be at this level or the system test level.

5. *System testing:*   Involves verifying the compatibility and joint performance of the software and hardware components of the total system. This level includes acceptance and quality assurance testing and system performance demonstrations.

6. *Operational testing:*   Involves providing realistic data to the computing system in the operational environment, under full operational demands. These tests are referred to as *operational soaks* or *operational demonstrations.*

9.5   Software Performance in the Use Stage

Since software performance characteristics are still vague, a good percentage of performance is improved in the use stage. In fact many bugs are introduced as a result of poor documentation, poorly structured software, and vague specification. At least five areas are critical in the use stage:

- *Failure logs:*   To use lessons learned on future projects.

- *Failure management:*   To assure there is a closed-loop system for removing the bugs permanently.

- *Configuration control:*   Formal procedures to control changes made in software. The procedures assure that the basic structure of the software design is not changed and that all changes are accounted for.

- *Customer support:*   To assure quality of services.

- *Software maintenance*

The last item is a new and complex area. The broadest definition is offered by Morton:[19] "Maintenance is any change to an existing program." This view is helpful because the source of the problem is change, not the type of change. Morton's definition of maintenance includes:

- Major enhancements
- Bug fixes
- Local adaptation of vendor-supplied systems
- Installation of vendor-supplied systems
- Adaptation to environmental changes such as new hardware

For the purpose of configuration control, Morton also suggests following guidelines:

Any change in the functional capabilities or performance specifications is a requirements change.

Any other user-visible change (other than fixing a bug) is a change in the functional specifications.

Any change in the allocation of functions to software components or any change in interfaces between components is an architectural change.

Any change in the design of a module, which is limited to only one module, is a module design change.

Any change to the code which does not also change the module design is a coding change.

Any bug after testing requires two fixes, one in the system (at whatever levels may be appropriate) and one in the test.

Reference 20 contains additional information on software maintenance.

9.6   Topics for Student Projects and Theses

1   Differentiate between software quality and software reliability. Summarize various approaches. Suggest your own version if the current definitions are unsatisfactory.

2   Summarize the techniques for achieving software maintainability goals. If the present techniques are inadequate, suggest an alternative technique.

3   Write a short report on design approaches for fault tolerance for human errors.

4   Summarize reliability measurement models. Recommend a model (or models) and justify your selection.

5 Develop a model for predicting software reliability which can be applied before the software is coded.

6 Are the current software safety approaches adequate? If they are adequate, justify. Propose your own approach.

7 Explain your views on approaches to software logistics analysis.

8 Can FMEA be applied to software alone? If so, perform this analysis on a portion of a software.

9 Can fault-tree analysis be applied to software alone? Summarize your views with an example.

10 What kind of structured programming is best suited for software maintenance. Present your views.

References

1. U.S. Army Solicitation No. DAH CO6-85-R-0018.
2. Edward C. Soistman and Katherine B. Ragsdale, *Impact of Hardware/Software Faults on System Reliability*, Report No. OR 18, 173, Griffiths Air Force Base, Rome Air Development Center, Rome, N.Y., 1985.
3. *Defense System Software Development*, DoD-Std-2167A, 1986.
4. *System Safety Program Requirements*, Mil-Std-882B, Naval Publications and Forms Center, 5801 Tabor Ave., Philadelphia, PA. 19120
5. General Electric Co., *Software Engineering Handbook*, McGraw-Hill, New York, 1986.
6. Fred Brooks, "No Silver Bullet: Essence and Accidents of Software Engineering," *Computer*, April 1987.
7. *A Standard Classification for Software Errors, Faults and Failures*, IEEE Draft Standard No. P1044/D3, IEEE, New York, December 1987.
8. Robert C. Clardy, "Sneak Analysis: An Integrated Approach," *International System Safety Conference Proceedings*, 1977.
9. Chenoweth and Schulmeyer, "...," *Proceedings of IEEE Conference on Computer Assurance*, IEEE Computer Software Assurance Conference, COMPSAC 86, Washington, D.C., 1986.
10. *Software System Safety Handbook*, AFISC SSH 1-1, U.S. Air Force Inspection and Safety Center, Norton Air Force Base, Calif., 1985.
11. Nancy Laveson, "Safety Analysis Using Petri Nets," IEEE Transactions on Software Engineerings, 1986.
12. Steven Mattern, Confessions of a Modern-Day Software Safety Analysis, *Proceedings of the Ninth International System Safety Conference*, Long Beach, California, 1989. Group Meeting, June 1989.
13. A. L. Goel and Kayudina Okumoto, "A Time Dependent Error Rate Model for Software Reliability and Other Performance Measures," *IEEE Transactions on Reliability*, no. R-28, 1979, pp. 206–211.
14. Germain Sagols and Jean Luc Albin, "Reliability Models and Software Development, A Practical Approach," *Software Engineering: Practice and Experience*, E. Girard (ed.), Oxford Publishing, Oxford, England, 1984.
15. *Reliability Qualification and Production Approval Tests*, Mil-Std-781, Naval Publications and Forms Center, 5801 Tabor Avenue, Philadelphia, PA., 19120
16. Glenford Myers, *Software Reliability: Principles and Practice*, John Wiley and Sons, New York, 1976.

17. Proposed version of Mil-Std-882C, being revised at the time of this writing. Contact Harris Yeager, Air Force Systems Command, Andrews Air Force Base, MD 20334-5000.
18. Paul P. Howley, Jr., "A Comprehensive Software Testing Methodology," *IEEE Workshop on Software Engineering Standards*, San Francisco, Calif., 1983.
19. Richard P. Morton, "The Application of Software Development Standards to Software Maintenance Tasks," *IEEE Workshop on Software Engineering Standards*, 1983.
20. David H. Longstreet, *Software Maintenance and Computers*, IEEE Computer Society Press, New York, 1990.

Further Reading

Adrian, W. R., M. A. Branstad, and J. C. Cherniavsky, "Validation, Verification, and Testing of Computer Software," Special Publication 500-75, National Bureau of Standards, Gaithersville, Md., February 1981.

Chikofsky, Elliot J., *Computer-Aided Software Engineering*, IEEE Computer Society Press, New York, 1988.

Chow, T. S., *Software Quality Assurance: A Practical Approach*, IEEE Computer Society Press, New York, 1985.

Curtis, Bill, *Human Factors in Software Development*, IEEE Computer Society Press, New York, 1986.

Defense System Software Development Handbook, DoD-Hdbk-287, Naval Publications and Forms Centers, 5801 Tabor Avenue, Philadelphia, PA 19120.

Donahoo, J. D., and D. Swearinger, "A Review of Software Maintenance Technology," Report RADC-TR-80-13, Rome Air Development Center, Rome, New York, 1980.

Dunn, Robert, and Richard Ullman: *Quality Assurance for Computer Software*, McGraw-Hill, New York, 1981.

Freeman, Peter, *Software Reusability*, IEEE Computer Society Press, New York, 1987.

Glass, Robert L., *Checkout Techniques, Software Reliability Guidebook*, Prentice-Hall, Englewood Cliffs, N.J., 1979.

IEEE Standard Dictionary of Measures to Produce Reliable Software, IEEE Standard 982.1-1988, IEEE, New York, 1988.

IEEE Standard for Software Reviews and Audits, IEEE Standard 1028-1988, IEEE, New York, 1988.

McGowan, C. L., *Top-Down Structured Programming Techniques*, Petrocelli/Charter, New York, 1975.

Musa, J. D., A. Iannine, and Kayudina Okumotoko, *Software Reliability: Measurement, Prediction, Application*, McGraw-Hill, New York, 1987.

Myers, Glenford J., *The Art of Software Testing*, John Wiley and Sons, New York, 1979.

Reviews and Audits for Systems, Equipment, and Computer Programs, Mil-Std-1521B, Naval Publications and Forms Center, 5801 Tabor Avenue, Philadelphia, PA 19120.

Shooman, Martin, *Software Engineering—Design, Reliability, Management*, McGraw-Hill, New York, 1983.

Software Quality Evaluation, DoD-Std-2168, Naval Publications and Forms Center, 5801 Tabor Avenue, Philadelphia, PA 19120.

System Safety Program Requirements, Mil-Std-882, Naval Publications and Forms Center, 5801 Tabor Avenue, Philadelphia, PA 19120, 1984.

System Effectiveness

10.1 Introduction

There is a saying, "Behind every technical disaster is a management shortcoming." Quality experts say management is responsible for 85 percent of quality problems. Reliability experts say the cause of unreliability is that there is never enough budget and time for thoroughly analyzing designs. The fact that maintenance costs usually are high suggests that management should get strongly involved in integrating such issues. Product costs can be cut drastically if waste in manufacturing and excess field failure costs are avoided. Avoidance of these costs also results in lower downtime costs to the customer. With integration of assurance technologies, both the supplier and the customer benefit. That is the secret of market leadership.

Typical situations mentioned below highlight the dollars lost by the companies. If these dollars were saved, companies might not need loans from banks; more capital would be generated internally than a bank could offer. Companies spend tremendous costs for burn-in and screening tests instead of solving the problems. Everyone recognizes that it is cheaper to solve problems than to test and inspect products for them. A lack of teamwork and poor analysis prevents this from happening. Millions of dollars can be saved by avoiding unnecessary tests. Moreover, good tests are often overlooked because management is already burdened with test costs. Waste because of inadequate manufacturing processes falls into the same category. These costs in some companies amount to 10 to 20 percent of sales.

Many companies waste enormous amounts of money because of poor integration among the technical departments. In one company, design engineers conducted all kinds of tests to demonstrate high reliability and durability in the products, but the manufacturing department rarely evaluated with such thoroughness. The result was that many engineering dollars were put into good designs which the manufacturing department could never deliver. The problems are often known but no one seems to be accountable or has the time to coordinate.

Assurance technologies are not independent of each other. Information on reliability is used for maintainability, quality, and logistics analyses. Test programs are developed from various analyses. Almost every product requires tradeoffs between

- Reliability and maintainability
- Functional performance and reliability
- Number of functions and reliability (in a microprocessor, for example)
- Maintainability and maintenance
- Hardware and software
- Component tolerances and reliability
- Human factors and design cost
- Operability and reliability
- Operational readiness and inventory cost

It is management's responsibility to provide environment that encourages free discussion of these tradeoffs. This will require team effort in various analyses and the budget. But there is a potential payoff of $40 to $100 for every extra dollar in the budget.

10.2   System Effectiveness Principles

For a system to perform at expected levels, the reliability, maintainability, safety, and quality functions need to work in harmony. If the equipment breaks down, it must be repairable quickly so that the system is available to the user. From a customer's perspective, a system's availability is more important. For instance, a commuter used to taking a train at 6 a.m., and again at 6 p.m. expects the train to arrive at those hours. If the train breaks down every night but is repaired before 6 a.m. these failures are of little concern to the commuter. Availability can be improved either by not allowing the train to fail or by designing it in such a way that it can be repaired very fast. At the user level, availability replaces reliability. It is defined as the percent time the system is in working condition when the user needs it. Power companies usually measure their performance in terms of availability. One utility's rate increases are tied to availability; customers pay less if they experience higher than usual loss of power.

Availability can also be defined as a measure of the degree to which an item is in an operable and committable state at the start of the mission, when the mission is called for at an unknown (random) point in time.[1] The technical definition of availability in terms of mean time between failures (MTBF) and mean time to repair (MTTR) is covered in Chap. 4.

Availability is actually only part of the customer's needs. As the system grows in complexity, assurance technologies such as reliability, maintainability, and quality assurance begin to look like the pieces of a puzzle called *system effectiveness*. Engineers seem to apply these technologies to hardware system effectiveness alone or to software system effectiveness alone, but the truth is that performance also depends on the integration of software and hardware effectiveness along with that of the humans who operate the system. This broad picture is often overlooked. Someone who has the responsibility for the total picture (that means top management) should also be accountable for overall effectiveness. Table 10.1[2] gives a typical breakdown of hardware, software, and other causes of outages in software-controlled systems. The *Other* category includes several causes including humans.

The numbers indicate that software causes more outages than hardware causes. The largest percentage of failures (43.1 percent) is unaccounted for. In other words, no one is responsible for almost half the system problems. This makes accountability for system availability more important than any stand-alone technology.

The first element to be integrated in system effectiveness is availability. System availability can be modeled in terms of component availabilities A:

$$A_{\text{system}} = A_{\text{hardware}} \times A_{\text{software}} \times A_{\text{human}} \times A_{\text{interface}}$$

TABLE 10.1   System Outages in Minutes by Causes

Month	Hardware	Software	Other	Total
1	43	102	195	340
2	28	121	0	149
3	133	39	21	193
4	31	81	119	231
5	261	128	411	800
6	76	109	916	1101
7	284	44	7	335
8	0	209	0	209
9	17	181	46	244
10	13	158	34	205
11	3	107	36	146
12	154	100	48	302
Mean	86.9	114.9	152.8	354.6
Percentage	24.5	32.4	43.1	100.00

Here, A_{human} does not merely mean that a human is available to do the work; it means that a trained and alert human who knows what to do during emergencies is available. $A_{\text{interface}}$ refers to the proportion of time the interfaces between the hardware, software, and human are in the usable condition.

The next item to be integrated is dependability. The system is not acceptable if it is not dependable, and management therefore cannot trust it. It also has to be capable of performing missions correctly throughout the intended period. When these principles are integrated, the result is system effectiveness. Reference 3 contains models and their sources. The information below is derived from this reference.

System effectiveness may be defined[4] as "The probability that the system can successfully meet an operational demand within a given time when operated under specified conditions." For a single-shot device, the system effectiveness can be "The probability that the system (such as a parachute) will operate successfully when called upon to do so under specified conditions." The Mil-Std-721[1] definition is "A measure of the degree to which an item can be expected to achieve a set of specific mission requirements and which may be expressed as a function of availability, maintainability, dependability, and capability." The WSEIAC model (Weapon System Effectiveness Industry Advisory Committee) developed by the U.S. Air Force Systems Command seems to be accepted by most practitioners. The U.S. Navy has also developed a model based on the WSEIAC model, as has the U.S. Army. These models are presented in Fig. 10.1. For more details, see Refs. 5, 6, and 7.

Some features shown in these models were developed for special applications. Others are unique to the organization. However, two commonly used terms can be defined:

Capability: The probability that the system's designed performance level will allow it to meet mission demands successfully provided that the system is available and dependable (Mil-Std-721). Capability includes elements such as consistency of power supply, fidelity, and precision.

Dependability: A measure of the item's operating condition at one or more operating points during the mission, including the effects of reliability, maintainability, and survivability, given the item's condition at the start of the mission (Mil-Std-721). Dependability includes safety and ease of maintenance.

10.3 Implementing the Programs

It is obvious that, if there is no plan to carry out certain activities, there will be no budget for them. Without a budget the tasks will not

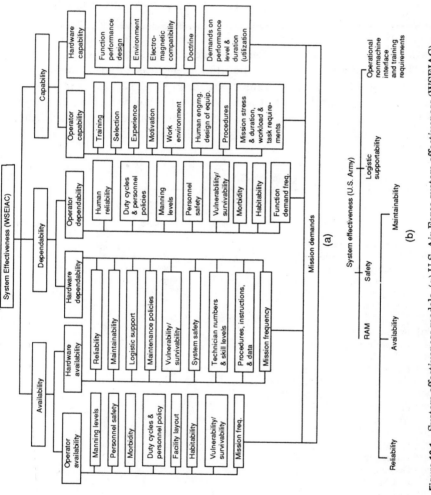

Figure 10.1 System effectiveness models. (*a*) U.S. Air Force system effectiveness (WSEIAC) model; (*b*) U.S. Navy reliability, availability, and maintenance (RAM) and system effectiveness model.

be done or will not be done on time. This affects engineering morale and productivity. Many so-called total quality management programs have failed because there were no real resources behind them.

Assurance programs.   Table 10.2 presents programs for reliability (Mil-Std-785)[8] and Table 10.3 presents programs for maintainability (Mil-Std-470).[9] Chapter 5 includes programs for system safety. Similar programs can be constructed for quality assurance from DoD directive 5000.51-G[1] or from Mil-Q-9858,[10] and for logistic engineering from Mil-Std-1388A.[11] These programs are not exhaustive; if a certain task is of proper value, it should be added to the program.

The reliability and maintainability programs have some similar tasks. A good program manager will group the similar and related tasks together and have them analyzed together. For example, FMECA is used for reliability as well as maintainability, but for different objectives. About 80 percent of the analysis will be common.

TABLE 10.2   Reliability Program (Mil-Std-785)

Task	Concept	Validation	FSD	Production
		Program phase		
101 Reliability Program Plan	x	x		
102 Monitoring Subcontractors		x	x	x
103 Program Reviews		x	x	x
104 Failure Reporting, Analysis, and Corrective Action System (FRACAS)		x	x	x
105 Failure Review Board (FRB)		x	x	x
201 Reliability Modeling	x	x		
202 Reliability Allocation	x	x		
203 Reliability Predictions	x	x		
204 Failure Modes, Effects, and Criticality Analysis (FMECA)	x	x		
205 Sneak Circuit Analysis (SCA)		x		
206 Parts Circuits Tolerance Analysis		x		
207 Parts Program		x	x	
208 Reliability Critical Items		x	x	
209 Effects on Storage, Handling, Packaging, Transportation, and Maintenance		x		
301 Environmental Stress Screening		x	x	
302 Reliability Growth Testing		x		
303 Reliability Qualification Test Program		x	x	
304 Production Reliability Acceptance Tests				x

TABLE 10.3   Maintainability Program (Mil-Std-470)

Task	Concept	Valida-tion	FSD	Produc-tion	Oper-ating system develop-ment
				Program phase	
101 Maintainability Program Plan		x	x	x	x
102 Monitor/Control of Subcontrac-tors and Vendors			x	x	
103 Program Reviews	x	x	x	x	
104 Data Collection, Analysis, and Corrective Action System			x	x	
201 Maintainability Modeling	x		x		
202 Maintainability Allocations	x		x		
203 Maintainability Predictions			x		
204 Failure Modes and Effects Analysis (FMEA) Maintainabil-ity Information			x		
205 Maintainability Analysis	x	x			
206 Maintainability Design Criteria			x		
207 Preparation of Inputs to De-tailed Maintenance Plan and Logistics Support Analysis (LSA)			x		
208 Maintainability Demonstration (MD)			x		

The results of the FMECA are also necessary for logistics analysis. It will be highly productive if the design engineer, reliability engineer, logistics engineer, and maintainability engineer conduct the analysis as a team. More and more companies are finding that the safety engineer should also be a team member.

A company cannot implement the complete intent of these programs on every product. The programs must be tailored for the project. Tailoring is explained by the following illustration: Consider a large dining table, with 14 legs. The table may be safe and reliable, but it may not be cost-effective. With four legs, however, it may provide the desired level of safety and reliability at a reasonable cost. *Tailoring* means picking out the right tasks with due consideration to cost and schedule constraints. It also means that, if a new task will make the product much safer, it should be added to the program. If the dining table had only three legs, it might be wise to add a fourth leg.

A major aspect of implementing the tasks is to perform them efficiently by using the principles of concurrent engineering. Figure 10.2 gives an example of tasks and their concurrent schedules.

Tasks	Design phase	Manufacturing phase	Test phase	Use phase
1. Management				
a. Program plan	▌			
b. Budgets	▌			
2. General				
a. Preliminary design review	▌			
b. Intermediate design reviews	▌			
c. Critical design reviews	▌			
3. Reliability				
a. Reliability prediction allocation and modeling	▌			
b. Failure mode and effects analysis	▌			
c. Worst case analysis	▌—▌			
d. Parts tolerance analysis	▌—▌			
e. Sneak circuit analysis	▌			
f. Parts program	▌——▌			
g. Analysis of storage, handling, packaging, and transportation	▌——▌			
h. Reliability growth tests		▌——▌		
i. Reliability qualification tests		▌———▌		
j. Production and screening tests		▌————▌		
k. Process reliability		▌—▌		
l. Software reliability analysis	▌———▌			
m. Human factors analysis for reliability	▌			
4. Maintainability				
a. Maintainability analysis	▌—▌			
b. Maintainability modeling and procedures and allocations	▌—▌			
c. Failure mode and effects analysis for maintainability	▌—▌			
d. Testability analysis	▌——▌			
e. Human factors analysis for maintainability	▌—▌			
f. Maintainability demonstration				▌

5. System safety

a. System safety program plan
b. Preliminary hazard analysis
c. Failure mode and effects analysis for safety
d. Fault tree analysis
e. Subsystem hazard analysis
f. Operations and support hazard analysis
g. Maintenance engineering safety analysis
h. System hazard analysis
i. Safety testing
j. Software hazard analysis
k. Emergency drills
l. Safety training

6. Quality assurance

a. Quality function deployment
b. Process reliability and availability analysis
c. Design for robustness analysis
d. Procurement plans
e. Pilot run evaluation
f. Process control plans
g. Continuous improvements
h. Software quality analysis

7. Logistic support

a. Logistic support analysis
b. Incorporation of the results of
   reliability and maintainability analysis
c. Time line analysis
d. Logistics analysis for manufacturing
e. Tests for logistics
f. Reliability centered maintenance planning

Figure 10.2 Master schedule of concurrent engineering tasks. Note that the manufacturing and test activities start during the design phase.

10.4 Managing by Life-Cycle Costs

A consumer rarely buys an automobile because it has the lowest price. Instead, a prospective buyer subconsciously looks at the costs of repair, maintenance, and fuel consumption. Decisions are usually based on life-cycle costs. Managers face similar challenges in making investments which may not provide immediate benefits. Unfortunately, many times they are not convinced there are enough benefits. They overlook many hidden ones. For example, capital equipment may be bought without evaluation of downtime costs. Higher-priced equipment with enormous savings in downtime may be rejected.

The concept of life-cycle costs in making engineering decisions has found wide token acceptance, but very few are willing to take it seriously. Two big reasons are (1) the short-term profits are reduced because the investment does not yield immediate gains; Wall Street does not like that; and (2) the investment is made by the design department but the gains are enjoyed by other departments such as manufacturing, marketing, and maintenance, and by the customer. Smart companies which do take advantage of the benefits of the concept capture higher market shares and profits in the long run. Short-sighted companies sooner or later go out of business, unless they have aggressive new product programs.

Chapter 4 gives examples of analyzing life-cycle costs. Another example is presented below to emphasize that such an approach is not only meaningful for selecting engineering options but also for making higher-level decisions. The life-cycle costs model generally contains current costs and future costs, to the supplier as well as to the buyer. The best situation is when the costs are reduced for the supplier as well as for the customer.

Example 10.1   A product contained a plastic gear on a plastic shaft. The product failed because the gear moved on the shaft. This was the result of expansion and contraction caused by changing environments. The analysis showed that the problem could be permanently designed out if the gear and the shaft were molded as one part. Even though the change made logical sense, the solution could not be implemented. The argument was that, since the investment in the existing process was already recovered, why pay $2 million for a new process. The life-cycle cost analysis, however, surprised the company. It discovered that the investment of $2 million would produce savings of more than $20 million over the next 10 years in terms of reduced warranty and related costs. This is another example of how life-cycle savings can be shared by the supplier as well as the customer. The customer benefits, of course, would be in terms of reduced failures.

10.5 System Effectiveness Model

As pointed out earlier, system effectiveness is a function of availability, dependability, and capability. Therefore, a model for system effectiveness may be written as a combination of probabilities:

$$P(\text{SE}) = P(A) \times P(D) \times P(C)$$

where SE, A, D, and C indicate that the probabilities are those of system effectiveness, availability, dependability, and capability.

One more term can be added to the system effectiveness model: operational readiness. This accounts for readiness of operating personnel and materiels, as in the U.S. Navy model in Fig. 10.1. The literature does not define operational readiness clearly. But the author recommends including the readiness of the operating crew and of support services such as ambulance, fire control, and rescue squads. The picture of assurance technologies now seems complete!

10.6 Author's Recommendations

The author has been credited with turning around a major division of a large corporation, which seemed on the verge of going out of business but instead achieved the number one spot as market leader in quality in less than 2 years. This took place in 1974. For years, many members of the corporate elite refused to believe the principles the author applied to the turnaround—until Japan used them on a massive scale. Some principles may still seem controversial, but the author is willing to share with the reader the truth, and nothing but the truth. The principles are embodied in the eight commandments of good product assurance management:

1. You shall always assess competitors' weaknesses and strengths to plan where you want to be.

2. You shall always make sure all the specifications are thoroughly reviewed for reliability, maintainability, safety, quality, and logistics support before detailed design work starts.

3. You shall minimize life-cycle costs by designing assurance features into the products and, as much as possible, not depend on inspection and testing.

4. You shall use concurrent engineering principles to make products faster, better, and cheaper.

5. You shall always review the manufacturing process as thoroughly as the product design before the process is installed.

6. You shall not use control charts in manufacturing until you have demonstrated that the problem cannot be eliminated completely. All inspection, including control charts, should be eliminated as soon as possible.

7. You shall not perform burn-in and screening on 100 percent of de-

vices unless you have demonstrated that the causes of failures cannot be eliminated.

8. You shall try to build customer trust and increase market share by eliminating causes of field failures.

10.7 Topics for Student Projects and Theses

1 Develop a model for system effectiveness other than the one described in this chapter. Include operational readiness.

2 Suggest your approach for an operational readiness model.

3 Suggest your approach to integration of assurance technologies. Consider cost, schedule, and performance in your approach.

4 Search for the best life-cycle model. Suggest your approach.

5 Discuss the role of management in system effectiveness.

References

1. *Definitions of Effectiveness Terms for Reliability, Maintainability, Human Factors, and Safety*, Mil-Std-721,
2. DoD Directive 5000.51-G.
3. F. A. Tillman, C. L. Hwang, and Way Kuo, "System Effectiveness Models: An Annotated Bibliography," *IEEE Transactions on Reliability*, no. R-29, October 1980.
4. ARINC Research Corp., *Reliability Engineering*, Prentice-Hall, Englewood Cliffs, N.J., 1964.
5. *Weapon Systems Effectiveness Industry Advisory Committee*, AFSC-TR-65, vols I, II, and III., U.S. Air Force Systems Command, Andrews Air Force Base, Md., 1965.
6. *Naval System Effectiveness Manual*, Technical Document 251 (NELC/TD251), Naval Electronics Laboratory Center, San Diego, 1973.
7. *Product Assurance—Army Materiel Reliability, Availability, and Maintainability*, AR 702-3, Department of the Army, Washington, D.C., 1973, revised 1976.
8. *Reliability Program for Systems and Equipment Development and Production*, Mil-Std-785,.
9. *Maintainability Program Requirements*, Mil-Std-470.
10. *Quality Program Requirements*, Mil-Q-9858.
11. *Logistic Support Analysis*, Mil-Std-1388, and *Logistic Support Analysis Records*, Mil-Std-1388-2A.

Further Reading

Arsenault, J. E., and J. A. Roberts (eds.), *Reliability and Maintainability of Electronic Systems*, Computer Science Press, Potomac, Md., 1980.
Electronic Reliability Design Handbook, Mil-Hdbk-338,
Proceedings of the Reliability and Maintainability Conference, annual publication, IEEE, New York.

Appendix

TABLE A   Standard Normal Distributions

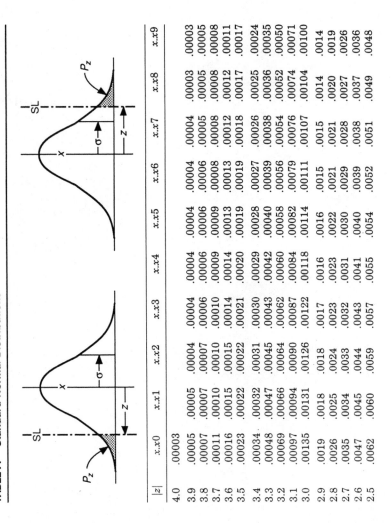

|z|	x.x0	x.x1	x.x2	x.x3	x.x4	x.x5	x.x6	x.x7	x.x8	x.x9
4.0	.00003									
3.9	.00005	.00005	.00004	.00004	.00004	.00004	.00004	.00004	.00003	.00003
3.8	.00007	.00007	.00007	.00006	.00006	.00006	.00006	.00005	.00005	.00005
3.7	.00011	.00010	.00010	.00010	.00009	.00009	.00008	.00008	.00008	.00008
3.6	.00016	.00015	.00015	.00014	.00014	.00013	.00013	.00012	.00012	.00011
3.5	.00023	.00022	.00022	.00021	.00020	.00019	.00019	.00018	.00017	.00017
3.4	.00034	.00032	.00031	.00030	.00029	.00028	.00027	.00026	.00025	.00024
3.3	.00048	.00047	.00045	.00043	.00042	.00040	.00039	.00038	.00036	.00035
3.2	.00069	.00066	.00064	.00062	.00060	.00058	.00056	.00054	.00052	.00050
3.1	.00097	.00094	.00090	.00087	.00084	.00082	.00079	.00076	.00074	.00071
3.0	.00135	.00131	.00126	.00122	.00118	.00114	.00111	.00107	.00104	.00100
2.9	.0019	.0018	.0018	.0017	.0016	.0016	.0015	.0015	.0014	.0014
2.8	.0026	.0025	.0024	.0023	.0023	.0022	.0021	.0021	.0020	.0019
2.7	.0035	.0034	.0033	.0032	.0031	.0030	.0029	.0028	.0027	.0026
2.6	.0047	.0045	.0044	.0043	.0041	.0040	.0039	.0038	.0037	.0036
2.5	.0062	.0060	.0059	.0057	.0055	.0054	.0052	.0051	.0049	.0048

z										
2.4	.0064	.0066	.0068	.0069	.0071	.0073	.0075	.0078	.0080	.0082
2.3	.0084	.0087	.0089	.0091	.0094	.0096	.0099	.0102	.0104	.0107
2.2	.0110	.0113	.0116	.0119	.0122	.0125	.0129	.0132	.0136	.0139
2.1	.0143	.0146	.0150	.0154	.0158	.0162	.0166	.0170	.0174	.0179
2.0	.0183	.0188	.0192	.0197	.0202	.0207	.0212	.0217	.0222	.0228
1.9	.0233	.0239	.0244	.0250	.0256	.0262	.0268	.0274	.0281	.0287
1.8	.0294	.0301	.0307	.0314	.0322	.0329	.0336	.0344	.0351	.0359
1.7	.0367	.0375	.0384	.0392	.0401	.0409	.0418	.0427	.0436	.0446
1.6	.0455	.0464	.0475	.0485	.0495	.0505	.0516	.0526	.0537	.0548
1.5	.0559	.0571	.0582	.0594	.0606	.0618	.0630	.0643	.0655	.0668
1.4	.0681	.0694	.0708	.0721	.0735	.0749	.0764	.0778	.0793	.0808
1.3	.0823	.0838	.0853	.0869	.0885	.0901	.0918	.0934	.0951	.0968
1.2	.0985	.1003	.1020	.1038	.1056	.1075	.1093	.1112	.1131	.1151
1.1	.1170	.1190	.1210	.1230	.1251	.1271	.1292	.1314	.1335	.1357
1.0	.1379	.1401	.1423	.1446	.1469	.1492	.1515	.1539	.1562	.1587
0.9	.1611	.1635	.1660	.1685	.1711	.1736	.1762	.1788	.1814	.1841
0.8	.1867	.1894	.1922	.1949	.1977	.2005	.2033	.2061	.2090	.2119
0.7	.2148	.2177	.2206	.2236	.2266	.2297	.2327	.2358	.2389	.2420
0.6	.2451	.2483	.2514	.2546	.2578	.2611	.2643	.2676	.2709	.2743
0.5	.2776	.2810	.2843	.2877	.2912	.2946	.2981	.3015	.3050	.3085
0.4	.3121	.3156	.3192	.3228	.3264	.3300	.3336	.3372	.3409	.3446
0.3	.3483	.3520	.3557	.3594	.3632	.3669	.3707	.3745	.3783	.3821
0.2	.3859	.3897	.3936	.3974	.4013	.4052	.4090	.4129	.4168	.4207
0.1	.4247	.4286	.4325	.4364	.4404	.4443	.4483	.4522	.4562	.4602
0.0	.4641	.4681	.4721	.4761	.4801	.4840	.4880	.4920	.4960	.5000

P_z = the proportion of process output beyond a particular value of interest (such as a specification limit) that is z standard deviation units away from the process average (for a process that is in statistical control and is normally distributed). For example, if $z = 2.17$, $P_z = .0150$ or 1.5%. In any actual situation, this proportion is only approximate.

TABLE B Constants and Formulas for Control Charts

	$\bar{X}$ and R charts*					$\bar{X}$ and s charts*			
	Chart for averages ($\bar{X}$)	Chart for ranges (R)			Chart for averages ($\bar{X}$)	Chart for standard deviations s			
	Factors for control limits	Divisors for estimate of standard deviation	Factors for control limits		Factors for control limits	Divisors for estimate of standard deviation	Factors for control limits		
Sub-group size n	A_2	d_2	D_3	D_4	A_3	c_4	B_3	B_4
---	---	---	---	---	---	---	---	---
2	1.880	1.128	—	3.267	2.659	0.7979	—	3.267
3	1.023	1.693	—	2.574	1.954	0.8862	—	2.568
4	0.729	2.059	—	2.282	1.628	0.9213	—	2.266
5	0.577	2.326	—	2.114	1.427	0.9400	—	2.089
6	0.483	2.534	—	2.004	1.287	0.9515	0.030	1.970
7	0.419	2.704	0.076	1.924	1.182	0.9594	0.118	1.882
8	0.373	2.847	0.136	1.864	1.099	0.9650	0.185	1.815
9	0.337	2.970	0.184	1.816	1.032	0.9693	0.239	1.761
10	0.308	3.078	0.223	1.777	0.975	0.9727	0.284	1.716
11	0.285	3.173	0.256	1.744	0.927	0.9754	0.321	1.679
12	0.266	3.258	0.283	1.717	0.886	0.9776	0.354	1.646
13	0.249	3.336	0.307	1.693	0.850	0.9794	0.382	1.618
14	0.235	3.407	0.328	1.672	0.817	0.9810	0.406	1.594
15	0.223	3.472	0.347	1.653	0.789	0.9823	0.428	1.572
16	0.212	3.532	0.363	1.637	0.763	0.9835	0.448	1.552
17	0.203	3.588	0.378	1.622	0.739	0.9845	0.466	1.534
18	0.194	3.640	0.391	1.608	0.718	0.9854	0.482	1.518
19	0.187	3.689	0.403	1.597	0.698	0.9862	0.497	1.503
20	0.180	3.735	0.415	1.585	0.680	0.9869	0.510	1.490

21	0.173	3.778	0.425	1.575	0.663	0.9876	0.523	1.477
22	0.167	3.819	0.434	1.566	0.647	0.9882	0.534	1.466
23	0.162	3.858	0.443	1.557	0.633	0.9887	0.545	1.455
24	0.157	3.895	0.451	1.548	0.619	0.9892	0.555	1.445
25	0.153	3.931	0.459	1.541	0.606	0.9896	0.565	1.435

$$\text{UCL}_{\bar{X}}, \text{LCL}_{\bar{X}} = \bar{\bar{X}} \pm A_2\bar{R}$$
$$\text{UCL}_R = D_4\bar{R}$$
$$\text{LCL}_R = D_3\bar{R}$$
$$\hat{\sigma} = \bar{R}/d_2$$

$$\text{UCL}_{\bar{X}}, \text{LCL}_{\bar{X}} = \bar{\bar{X}} \pm A_3\bar{s}$$
$$\text{UCL}_s = B_4\bar{s}$$
$$\text{LCL}_s = B_3\bar{s}$$
$$\hat{\sigma} = \bar{s}/c_4$$

*From ASTM publication STP-15D, *Manual on the Presentation of Data and Control Chart Analysis*, 1976; pp. 134–136. Copyright ASTM, 1916 Race Street, Philadelphia, Pa. 19103. Reprinted with permission.

TABLE C   Percentage Points of the χ^2 Distribution

Table of $\chi_{\alpha;\nu}$—the 100 α percentage point of the χ^2 distribution for ν degrees of freedom

$\chi_{\alpha;\nu}$

ν \ α	.995	.99	.98	.975	.95	.90	.80	.75	.70	.50
1	$.0^4393$	$.0^3157$	$.0^3628$	$.0^3982$	.00393	.0158	.0642	.102	.148	.455
2	.0100	.0201	.0404	.0506	.103	.211	.446	.575	.713	1.386
3	.0717	.115	.185	.216	.352	.584	1.005	1.213	1.424	2.366
4	.207	.297	.429	.484	.711	1.064	1.649	1.923	2.195	3.357
5	.412	.554	.752	.831	1.145	1.610	2.343	2.675	3.000	4.351
6	.676	.872	1.134	1.237	1.635	2.204	3.070	3.455	3.828	5.348
7	.989	1.239	1.564	1.690	2.167	2.833	3.822	4.255	4.671	6.346
8	1.344	1.646	2.032	2.180	2.733	3.490	4.594	5.071	5.527	7.344
9	1.735	2.088	2.532	2.700	3.325	4.168	5.380	5.899	6.393	8.343
10	2.156	2.558	3.059	3.247	3.940	4.865	6.179	6.737	7.267	9.342
11	2.603	3.053	3.609	3.816	4.575	5.578	6.989	7.584	8.148	10.341
12	3.074	3.571	4.178	4.404	5.226	6.304	7.807	8.438	9.034	11.340
13	3.565	4.107	4.765	5.009	5.892	7.042	8.634	9.299	9.926	12.340
14	4.075	4.660	5.368	5.629	6.571	7.790	9.467	10.165	10.821	13.339
15	4.601	5.229	5.985	6.262	7.261	8.547	10.307	11.036	11.721	14.339
16	5.142	5.812	6.614	6.908	7.962	9.312	11.152	11.912	12.624	15.338
17	5.697	6.408	7.255	7.564	8.672	10.085	12.002	12.792	13.531	16.338
18	6.265	7.015	7.906	8.231	9.390	10.865	12.857	13.675	14.440	17.338
19	6.844	7.633	8.567	8.907	10.117	11.651	13.716	14.562	15.352	18.338
20	7.434	8.260	9.237	9.591	10.851	12.443	14.578	15.452	16.266	19.337
21	8.034	8.897	9.915	10.283	11.591	13.240	15.445	16.344	17.182	20.337
22	8.643	9.542	10.600	10.982	12.338	14.041	16.314	17.240	18.101	21.337
23	9.260	10.196	11.293	11.688	13.091	14.848	17.187	18.137	19.021	22.337
24	9.886	10.856	11.992	12.401	13.848	15.659	18.062	19.037	19.943	23.337
25	10.520	11.524	12.697	13.120	14.611	16.473	18.940	19.939	20.867	24.337
26	11.160	12.198	13.409	13.844	15.379	17.292	19.820	20.843	21.792	25.336
27	11.808	12.879	14.125	14.573	16.151	18.114	20.703	21.749	22.719	26.336
28	12.461	13.565	14.847	15.308	16.928	18.939	21.588	22.657	23.647	27.336
29	13.121	14.256	15.574	16.047	17.708	19.768	22.475	23.567	24.577	28.336
30	13.787	14.953	16.306	16.791	18.493	20.599	23.364	24.478	25.508	29.336

SOURCE: *RELIABILITY HANDBOOK*, AMCP 702-3, HEADQUARTERS, U. S. ARMY MATERIAL COMMAND, WASHINGTON D.C., 1968.

TABLE C   Percentage Points of χ^2 Distribution (*Continued*)

.30	.25	.20	.10	.05	.025	.02	.01	.005	.001	α ν
1.074	1.323	1.642	2.706	3.841	5.024	5.412	6.635	7.879	10.827	1
2.408	2.773	3.219	4.605	5.991	7.378	7.824	9.210	10.597	13.815	2
3.665	4.108	4.642	6.251	7.815	9.348	9.837	11.345	12.838	16.268	3
4.878	5.385	5.989	7.779	9.488	11.143	11.668	13.277	14.860	18.465	4
6.064	6.626	7.289	9.236	11.070	12.832	13.388	15.086	16.750	20.517	5
7.231	7.841	8.558	10.645	12.592	14.449	15.033	16.812	18.548	22.457	6
8.383	9.037	9.803	12.017	14.067	16.013	16.622	18.475	20.278	24.322	7
9.524	10.219	11.030	13.362	15.507	17.535	18.168	20.090	21.955	26.125	8
10.656	11.389	12.242	14.684	16.919	19.023	19.679	21.666	23.589	27.877	9
11.781	12.549	13.442	15.987	18.307	20.483	21.161	23.209	25.188	29.588	10
12.899	13.701	14.631	17.275	19.675	21.920	22.618	24.725	26.757	31.264	11
14.011	14.845	15.812	18.549	21.026	23.337	24.054	26.217	28.300	32.909	12
15.119	15.984	16.985	19.812	22.362	24.736	25.472	27.688	29.819	34.528	13
16.222	17.117	18.151	21.064	23.685	26.119	26.873	29.141	31.319	36.123	14
17.322	18.245	19.311	22.307	24.996	27.488	28.259	30.578	32.801	37.697	15
18.418	19.369	20.465	23.542	26.296	28.845	29.633	32.000	34.267	39.252	16
19.511	20.489	21.615	24.769	27.587	30.191	30.995	33.409	35.718	40.790	17
20.601	21.605	22.760	25.989	28.869	31.526	32.346	34.805	37.156	42.312	18
21.689	22.718	23.900	27.204	30.144	32.852	33.687	36.191	38.582	43.820	19
22.775	23.828	25.038	28.412	31.410	34.170	35.020	37.566	39.997	45.315	20
23.858	24.935	26.171	29.615	32.671	35.479	36.343	38.932	41.401	46.797	21
24.939	26.039	27.301	30.813	33.924	36.781	37.659	40.289	42.796	48.268	22
26.018	27.141	28.429	32.007	35.172	38.076	38.968	41.638	44.181	49.728	23
27.096	28.241	29.553	33.196	36.415	39.364	40.270	42.980	45.558	51.179	24
28.172	29.339	30.675	34.382	37.652	40.646	41.566	44.314	46.928	52.620	25
29.246	30.434	31.795	35.563	38.885	41.923	42.856	45.642	48.290	54.052	26
30.319	31.528	32.912	36.741	40.113	43.194	44.140	46.963	49.645	55.476	27
31.391	32.620	34.027	37.916	41.337	44.461	45.419	48.278	50.993	56.893	28
32.461	33.711	35.139	39.087	42.557	45.722	46.693	49.588	52.336	58.302	29
33.530	34.800	36.250	40.256	43.773	46.979	47.962	50.892	53.672	59.703	30

TABLE D   5, Median, and 95 Percent Ranks for Various Sample Sizes

5 Percent Ranks

j	\	\	\	\	\	\	\	\	Sample size, n											
	1	2	3	4	5	6	7	8	9	10	11	12	13	14	15	16	17	18	19	20
1	5.000	2.532	1.695	1.274	1.021	0.851	0.730	0.639	0.568	0.512	0.465	0.426	0.394	0.366	0.341	0.320	0.301	0.285	0.270	0.256
2		22.361	13.535	9.761	7.644	6.285	5.337	4.639	4.102	3.677	3.332	3.046	2.805	2.600	2.423	2.268	2.132	2.011	1.903	1.806
3			36.840	24.860	18.925	15.316	12.876	11.111	9.775	8.726	7.882	7.187	6.605	6.110	5.685	5.315	4.990	4.702	4.446	4.217
4				47.237	34.259	27.134	22.532	19.290	16.875	15.003	13.507	12.285	11.267	10.405	9.666	9.025	8.464	7.969	7.529	7.135
5					54.928	41.820	34.126	28.924	25.137	22.244	19.958	18.102	16.566	15.272	14.166	13.211	12.377	11.643	10.991	10.408
6						60.696	47.930	40.031	34.494	30.354	27.125	24.530	22.395	20.607	19.086	17.777	16.636	15.634	14.747	13.955
7							65.184	52.932	45.036	39.338	34.981	31.524	28.705	26.358	24.373	22.669	21.191	19.895	18.750	17.731
8								68.766	57.086	49.310	43.563	39.086	35.480	32.503	29.999	27.860	26.011	24.396	22.972	21.707
9									71.687	60.584	52.991	47.267	42.738	39.041	35.956	33.337	31.083	29.120	27.395	25.865
10										74.113	63.564	56.189	50.535	45.999	42.256	39.101	36.401	34.060	32.009	30.195
11											76.160	66.132	58.990	53.434	48.925	45.165	41.970	39.215	36.811	34.693
12												77.908	68.366	61.461	56.022	51.560	47.808	44.595	41.806	39.358
13													79.418	70.327	63.656	58.343	53.945	50.217	47.003	44.197
14														80.736	72.060	65.617	60.436	56.112	52.420	49.218
15															81.896	73.604	67.381	62.332	58.088	54.442
16																82.925	74.988	68.974	64.057	59.897
17																	83.843	76.234	70.420	65.634
18																		84.668	77.363	71.738
19																			85.413	78.389
20																				86.089

Median Ranks

j	1	2	3	4	5	6	7	8	9	10	11	12	13	14	15	16	17	18	19	20
												Sample Size, n								
1	50.000	29.289	20.630	15.910	12.945	10.910	9.428	8.300	7.412	6.697	6.107	5.613	5.192	4.830	4.516	4.240	3.995	3.778	3.582	3.406
2		70.711	50.000	38.573	31.381	26.445	22.849	20.113	17.962	16.226	14.796	13.598	12.579	11.702	10.940	10.270	9.678	9.151	8.677	8.251
3			79.370	61.427	50.000	42.141	36.412	32.052	28.624	25.857	23.578	21.669	20.045	18.647	17.432	16.365	15.422	14.581	13.827	13.147
4				84.090	68.619	57.859	50.000	44.015	39.308	35.510	32.380	29.758	27.528	25.608	23.939	22.474	21.178	20.024	18.988	18.055
5					87.055	73.555	63.588	55.984	50.000	45.169	41.189	37.853	35.016	32.575	30.452	28.589	26.940	25.471	24.154	22.967
6						89.090	77.151	67.948	60.691	54.831	50.000	45.951	42.508	39.544	36.967	34.705	32.704	30.921	29.322	27.880
7							90.572	79.887	71.376	64.490	58.811	54.049	50.000	46.515	43.483	40.823	38.469	36.371	34.491	32.795
8								91.700	82.038	74.142	67.620	62.147	57.492	53.485	50.000	46.941	44.234	41.823	39.660	37.710
9									92.587	83.774	76.421	70.242	64.984	60.456	56.517	53.059	50.000	47.274	44.830	42.626
10										93.303	85.204	78.331	72.472	67.425	63.033	59.177	55.766	52.726	50.000	47.542
11											93.893	86.402	79.955	74.392	69.548	65.295	61.531	58.177	55.170	52.458
12												94.387	87.421	81.353	76.061	71.411	67.296	63.629	60.340	57.374
13													94.808	88.298	82.568	77.525	73.060	69.079	65.509	62.289
14														95.169	89.060	83.635	78.821	74.529	70.678	67.205
15															95.484	89.730	84.578	79.976	75.846	72.119
16																95.760	89.730	85.419	81.011	77.033
17																	96.005	90.849	86.173	81.945
18																		96.222	91.322	86.853
19																			96.418	91.749
20																				96.594

95 Percent Ranks

j	1	2	3	4	5	6	7	8	9	Sample size 10	11	12	13	14	15	16	17	18	19	20
1	95.000	77.639	63.160	52.713	45.072	39.304	34.816	31.234	28.313	25.887	23.840	22.092	20.582	19.264	18.104	17.075	16.157	15.332	14.587	13.911
2		97.468	86.468	75.139	65.741	58.180	52.070	47.068	42.914	39.416	36.436	33.868	31.634	29.673	27.940	26.396	25.012	23.766	22.637	21.611
3			98.305	90.239	81.075	72.866	65.874	59.969	54.964	50.690	47.009	43.811	41.010	38.539	36.344	34.383	32.619	31.026	29.580	28.262
4				98.726	92.356	84.684	77.468	71.076	65.506	60.662	56.437	52.733	49.465	46.566	43.978	41.657	39.564	37.668	35.943	34.366
5					98.979	93.715	87.124	80.710	74.863	69.646	65.019	60.914	57.262	54.000	51.075	48.440	46.055	43.888	41.912	40.103
6						99.149	94.662	88.889	83.125	77.756	72.875	68.476	64.520	60.928	57.744	54.835	52.192	49.783	47.580	45.558
7							99.270	95.361	90.225	84.997	80.042	75.470	71.295	67.497	64.043	60.899	58.029	55.404	52.997	50.782
8								99.361	95.898	91.274	86.492	81.898	77.604	73.641	70.001	66.663	63.599	60.784	58.194	55.803
9									99.432	96.323	92.118	87.715	83.434	79.393	75.627	72.140	68.917	65.940	63.188	60.641
10										99.488	96.668	92.813	88.733	84.728	80.913	77.331	73.989	70.880	67.991	65.307
11											99.535	96.954	93.395	89.595	85.834	82.223	78.809	75.604	72.605	69.805
12												99.573	97.195	93.890	90.334	86.789	83.364	80.105	77.028	74.135
13													99.606	97.400	94.315	90.975	87.623	84.366	81.250	78.293
14														99.634	97.577	94.685	91.535	88.357	85.253	82.269
15															99.659	97.732	95.010	92.030	89.009	86.045
16																99.680	97.868	95.297	92.471	89.592
17																	99.699	97.989	95.553	92.865
18																		99.715	98.097	95.783
19																			99.730	98.193
20																				99.744

Index

ABOUT THE AUTHOR

Dev G. Raheja, President of Technology Management, Inc., and Lecturer at the Graduate School of the University of Maryland, is a consultant and professor specializing in assurance technologies such as quality assurance, reliability engineering, and product safety. With over 25 years of experience in the public and private sectors, Dev Raheja has worked as chief engineer responsible for corporate reliability and product safety engineering at the RTE Corporation, as senior systems assurance management consultant at Booz-Allen & Hamilton, Inc., and has held similar positions at General Electric, Allis Chalmers, and the Standard Screw Company. He has received several awards, including the Austin Bonis Reliability Education Award from ASQC and the Scientific Achievement Award from the System Safety Society. He has conducted seminars in more than 30 countries and for many Fortune 500 companies in the United States. An active Fellow of the American Society of Quality Control (ASQC), Dev Raheja is the Future Technologies Editor of ASQC's *Reliability Review*. He is a resident of Laurel, Maryland.